Basic Electricity
& Practical Wiring

Hoerner

Introduction

This manual, *Basic Electricity and Practical Wiring*, is for teachers and students of vocational and technical programs, as well as industrial and adult educational programs. Commercial firms and other agencies may also find it useful for in-service education programs.

Essential subject matter about electricity, especially as it affects our lives, is presented in a logical unit format. Classroom and teacher-demonstrated laboratory exercises are found at the end of major units. Students can reinforce the material they study by completing the classroom exercises and participate by assisting the instructor to safely perform the demonstrations found in the laboratory exercises. The final laboratory exercise is a set of six wiring projects that provide students with **hands-on learning**.

Subject matter includes working with parallel and series circuits, installing 120 and 240 volt circuits, measuring electrical power and costs, switching for electrical conductors, grounding for safety, providing adequate wiring and overload protection, transmitting and distributing electricity, and maintaining the service entrance. The concepts gained from this subject matter will help students understand the electrical demonstrations presented by the teacher and other students. It will also help the students perform the recommended practical wiring skills in the manual, and perform wiring maintenance or construction skills.

Safety is an important part of this manual. Electricity is dangerous, and this manual repeatedly reinforces the respect on must develop for electricity as the student and the teacher work with both the principles of electrical circuits and the wiring of actual circuits. Much of the **National Electrical Code**, from a safety standpoint, is presented without having to glean Code provisions directly from other sources. Informational includes proper circuit grounding and fusing of the **ground-fault circuit interrupter (GFCI)**. This discussion will help students develop an understanding of safety principles and a respect for electricity, demonstrated by their attitudes and habits.

Materials found in the Appendix help make the manual a more effective instructional tool. This material includes Ohm's Law and wattage formulas, wiring handbook tables, symbol charts, a set of definitions of electrical terms, and other information.

Supplemental instructional materials, designed to make teaching electricity more meaningful and reduce the time the instructor has to spend preparing, are available from Hobar Publications. These items include electrical wiring practice panels—complete and ready to use for hands-on wiring exercises; tool kits containing the essential hand tools for wiring; specially prepared equipment for performing the teacher-demonstration laboratory exercises; and other related visual and instructional materials. Contact Hobar Publications for a complete catalog.

Author:
Harry J. Hoerner
Professor Emeritus
Western Illinois University
Macomb, Illinois

Editorial Assistants:

Thomas A. Hoerner
Professor Emeritus
Iowa State University
Ames, Iowa

W. Forrest Bear
Professor Emeritus
University of Minnesota
St. Paul, Minnesota

HOBAR PUBLICATIONS
A Division of Finney Company
8075 215th Street West
Lakeville, Minnesota 55044
Phone: (800) 846-7027   Web site: www.finney-hobar.com

ISBN 10: 0-913163-42-2
ISBN 13: 978-0-913163-42-9

Copyright © 2007 by Hobar Publications

Cover images:
Crossed Wires © Stian Iversen. Image from BigStockPhoto.com.
Lightening Strike © Paul Brian. Image from BigStockPhoto.com.
Idea © Gabriel Moisa. Image from BigStockPhoto.com.

All rights reserved. No part of this book covered by the copyrights hereon may be reproduced or copied in any form or by any means—graphic, electronic, or mechanical, including photocopying, taping, or information storage and retrieval systems—without written permission of the publisher.

First Printing	1977
Second Printing	1978
Third Printing	1980
Fourth Printing	1984
First Revision	1991
Second Revision	2000
Third Revision	2007

Table of Contents

Unit	Page
INTRODUCTION TO ELECTRICITY	3
PARALLEL AND SERIES CIRCUITS AND 120 AND 240 VOLT CIRCUITS	5

    Parallel Circuits, Series Circuits, Combination Series-Parallel Circuits, Ohm's Law, Using Meters to Measure Values on Parallel and Series Circuits, Types of Voltages.

    Classroom Exercise I--Parallel and Series Circuits and 120 and 240 Volt Circuits ... 11

    Laboratory Exercise I--Parallel and Series Circuits and 120 and 240 Volt Circuits ... 14

    Laboratory Exercise II--Ohm's Law As Related to Parallel Circuits ... 17

MEASURING ELECTRIC POWER AND COSTS ... 19

    Watts and Kilowatts, Methods of Determining Power, Computing Power Factor of a Motor, Watts and Paying for Them.

    Classroom Exercise II--Measuring Electric Power and Costs ... 24

    Laboratory Exercise III--Measuring Electric Power and Costs ... 26

SWITCHING FOR ELECTRICAL CIRCUITS ... 29

    Single-Pole Switching, Three-Way Switching, Four-Way Switching.

    Classroom Exercise III--Switching for Electrical Circuits ... 32

    Laboratory Exercise IV--Switching for Electrical Circuits ... 33

GROUNDING FOR SAFETY ... 35

    The Grounding System, The Grounding System Applied to 240 Volt Applications, The Ground-Fault Interrupter.

    Classroom Exercise IV--Grounding for Safety ... 38

    Laboratory Exercise V--Grounding for Safety ... 39

PROVIDING FOR ADEQUATE WIRING AND OVERLOAD CURRENT PROTECTION ... 41

    Wire Size, Conductors and Insulators, Ampacity, Voltage Drop, Providing Proper Overload Current Protection, Overload Current Protection for Motors, Overload Current Protection for Resistance Lighting and Heating, Characteristics of Overcurrent Devices, Residential and Farm Circuits.

    Classroom Exercise V--Providing Adequate Wiring and Overload Protection ... 45

    Laboratory Exercise VI--Providing Adequate Wiring and Overload Current Protection ... 47

TRANSMISSION AND DISTRIBUTION OF ELECTRICITY ... 51

    From Energy to Electricity, Generators, The Transfer of Electricity, Transformers.

    Classroom Exercise VI--Transmission and Distribution of Electricity ... 54

THE SERVICE ENTRANCE 55
  Overhead in Air, Underground, Types of Cables and Conductors
  for Service Entrances, The Entrance Panel, Selecting the
  Service Entrance Panel and Disconnect.

  Classroom Exercise VII--The Service Entrance 62

PRACTICAL WIRING LABORATORY EXERCISES 65

  Evaluation of Wiring Practice Exercises 71

  Basic Electricity and Practical Wiring Worksheets 72

APPENDIX 78

  Ohm's Law and Wattage Formulas 78

  Wiring Handbook Tables 79

  Definition of Electrical Terms 84

INTRODUCTION TO ELECTRICITY

We live in a world of **electricity.** Electricity provides us with sources of heat, light, energy to produce rotary motion in electric motors, movement for sensing devices, and for transportation. We are therefore, dependent on electricity for food, clothing, shelter, entertainment, and recreation.

The agribusiness world is equipped with lighting to permit the agricultural worker to complete tasks more efficiently. Electric motors drive feed mills, silo unloaders, air compressors, and conveyors and there are electric drills, power saws, and welders for making repairs. The horticulturalist uses electricity and automatic sensing devices to control lights at the proper times. Seeds and grains are sieved by electrically operated machines in an agricultural business. In fact, there is an almost endless array of electrical applications to home, farm, agribusiness, and the recreational activity we all enjoy.

Electricity is often considered as magic to the beginning student which promotes a certain amount of fear into the product and its application. Therefore, when devices as simple as a fuse blows, and we are without electrical service, panic occurs. How do we restore the electrical system? Do we call the electrician, or do we attempt to assess the problem ourselves and restore service? With a basic understanding of electrical principles, we can determine whether to complete an electrical task or call a qualified electrician.

The study of this manual will not train a skilled electrician; however, it will help develop an understanding of electricity and do away with some of the fears and superstitions about electricity and electrical components. Understanding and knowledge tend to erase fears.

No one has ever said that electricity is not dangerous. The power of electricity commonly found in our home is more than ample to cause death from electrical shock. Therefore, the safe use of electricity plays a major role in this manual. Again, if the fundamentals of practical wiring, installing circuits, and proper fusing are known and if dangerous situations can be recognized, there is a better chance of not getting shocked, having a fire, or having an inconvenient loss of electrical service.

In the future, homes and agribusinesses will be served by more, not fewer, electrically operated appliances and devices. Electric home heating is relatively new; and air conditioning for livestock and automatically controlled environments for growth of horticultural products are fast becoming commonplace.

A student needs to understand through study and application the practical aspects of electricity. This manual is devoted to helping the student achieve that goal. Interested students may be motivated to study advanced electrical theories and applications. The units in this manual may also influence students to enter an electrical career.

The teacher and student will have an adventure in electricity when completing the use of this manual. A better understanding will be achieved; and, improved manipulative skills will be learned that will be used throughout life.

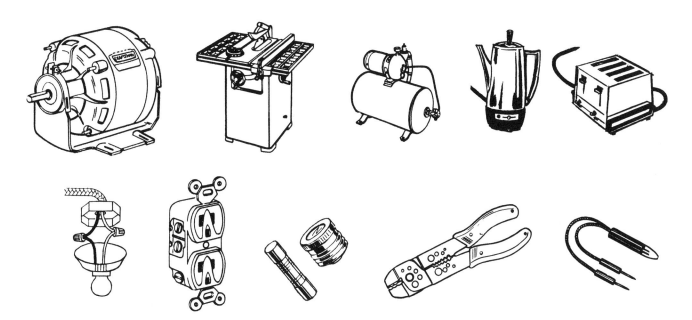

Fig. 1. Some Common Electrical Equipment and Appliances.

NOTES

PARALLEL AND SERIES CIRCUITS AND 120 AND 240 VOLT CIRCUITS

BASIC CIRCUITS

Electricity is primarily used in what is referred to as **circuits.** A circuit has two current-carrying conductors from the service entrance box to electrical fixtures. A typical residence (home) may have 8 to 24 circuits. The simple circuit shown in Figure 1 has the minimum components that any circuit must have, namely: (A) voltage source, (B) current-carrying conductors, and (C) an electrical load, such as an appliance, receptacle, or lamp.

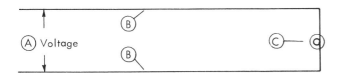

Fig. 1. A Simple Circuit.

A circuit may also have a directional device, such as a switch, which controls current flow to fixtures as illustrated in Figure 2 and identified by the letter (D).

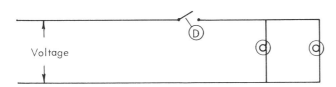

Fig. 2. A Circuit with a Switch Controlling Two Lamps.

PARALLEL CIRCUITS

Circuits can be classified into either **parallel** or **series** circuits. For the most part, circuits found in a home are **parallel circuits.**

Fig. 3. A Parallel Circuit.

The parallel circuit, Figure 3, is drawn as a circuit schematic. The upper part of the schematic has one current-carrying conductor labeled as black. The lower part of the schematic is a white current-carrying conductor. The two lamps (symbolized by ⊗ are **parallel** across the circuit. When the source voltage is energized, or turned **on**, we would expect the lamps to light because current will flow from black to white across each lamp. To this parallel circuit we could add other electrical fixtures if each is connected across the circuit, or between the black and the white conductors. An example of this is the parallel circuit in Figure 4, with three lamps across the black and white. One big advantage of the parallel circuit is that one lamp could go **out** or be disconnected and the remaining lamps would continue to operate.

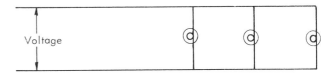

Fig. 4. A Parallel Circuit with Three Lamps.

SERIES CIRCUITS

Series circuits are constructed differently than parallel circuits. The main difference is that electrical fixtures, such as lamps, are in **series** with each other. Study the schematic in Figure 5 for the explanation. Note, as the current flows in the circuit it must follow a path through both lamps. However, if one lamp is disconnected, or if the lamp burns out, the current will stop flowing and the remaining lamp will also go out. This is the obvious disadvantage of any series circuit and is the main reason they are unpopular as circuits in residential and agricultural wiring applications. However, this **series** arrangement has advantages and is heavily used in certain appliances, electrical controls, motors and automotive wiring situations. Also, all switches are installed in a series relationship regardless of whether the circuit is a parallel or a series circuit, Figures 2 and 6.

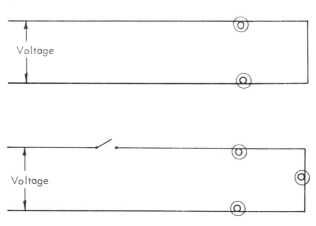

Fig. 6. A Circuit with Three Lamps in Series.

Three lamps are connected in series in Figure 6. Additional lamps could be added anywhere in the circuit if they are placed in a series relationship to the others, and not wired directly across the two leads of the source voltage.

COMBINATION SERIES-PARALLEL CIRCUITS

Both series and parallel circuits can be constructed within one total circuit and are called combination circuits. For example, Figure 7 illustrates lamp A as a parallel connection; whereas, lamps B and C are in series with each other and B and C are in parallel with A.

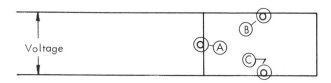

Fig. 7. A Combination Series-Parallel Circuit.

The use of combination circuits in home and agricultural wiring is practically void. However, they are used in internal wiring for electrical units, such as appliances, controls, sensing devices, automotive, and other pieces of equipment where applicable.

To gain a better understanding of circuits, it is necessary to put these circuits to work. It is also necessary to understand some of the basic laws governing circuits so that one knows what to expect when some common resistors are used with different circuit types.

OHM'S LAW

One of the most important laws of basic electricity is **Ohm's Law.** Dr. George Sim Ohm introduced this law to fellow German physicists in 1827. He stated that **volts equal amperes times resistance.** Letters or symbols are used to abbreviate these words and a definition of each follows:

- **V = Volts . . . The force or pressure of electricity.** In the past the letter E has been used to represent volts and the E is derived from the word **Electromotive force.** V is more acceptable today and is the metric symbol for volts.
- **A = Amperes . . . The rate of flow of electricity.** The letter I has previously been used to represent amperes; however, **A** is now more acceptable and is also the metric symbol.
- **R = Resistance . . . That which opposes the flow of electricity.** Resistance is measured in ohms. The Greek symbol omega (Ω) is used to express the amount of resistance in the metric system; however, **R** will be used for ease of presentation.

Ohm's Law can be expressed in three different ways as follows:

$$V = A \times R \qquad R = \frac{V}{A} \qquad A = \frac{V}{R}$$

Fig. 8. Ohm's Law Circle.

An Ohm's Law circle, Figure 8, is a convenient aid to use when working with Ohm's Law. It can be used by placing a finger over the letter for which a problem solution is desired. For example, if volts and resistance are known, place the finger over A, and the V appears above the line and R under the line as in Figure 9. Therefore, A is equal to V (voltage) divided by R (resistance). Any of the three factors can be determined if two of the three values are known. For example, if voltage is 120, and resistance is 60, then the amperes flowing in the circuit would be 2, because

$$A = \frac{V}{R} = \frac{120}{60} = 2.$$

In another example, if amperes equal 2.4 and volts equal 240, then the resistance is 100, because:

$$R = \frac{V}{A} = \frac{240}{2.4} = 100 \text{ ohms.}$$

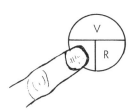

Fig. 9. Using the Ohm's Law Circle.

An electrical resistor such as a light bulb has a relatively constant resistance. However, most of these devices will not have the resistance stated on them, but the resistance can be calculated if we use another basic electrical formula. The basic unit of electrical power is the watt, and watts (W) equals amperes (A) times volts (V). It can also be illustrated in the wattage circle, Figure 10, and used similarly to the Ohm's Law Circle. Another circle, presenting a number of formulas for both Ohm's Law and wattage, is found in the Appendix.

Fig. 10. Wattage Circle.

Most electrical appliances (water heaters, fry skillets, toasters) and light bulbs commonly list either (1) wattage and amperes, or (2) amperes and voltage, or (3) wattage and voltage. Therefore, we can compute the resistance of a load by using both the wattage formula and Ohm's Law formula. An example would be to determine the resistance of a 60 watt light bulb stamped for 120 volts. Using the wattage formula, amperes equals watts divided by voltage, or:

$$A = \frac{60}{120} = 0.5$$

Once amperes are known, then substituting amperes and voltage into the Ohm's Law formula, resistance can be calculated. In this case, resistance equals voltage divided by amperes, or:

$$R = \frac{120}{0.5} = 240 \text{ Ohms}$$

USING METERS TO MEASURE ELECTRICAL VALUES ON PARALLEL AND SERIES CIRCUITS

Four meters are commonly used to measure electrical activity. These are the **voltmeter, ammeter, ohmmeter, and wattmeter**. Wattmeters and their use will be discussed in the next unit.

Voltmeters are used to measure the electrical potential between two points in a circuit. A voltmeter in a parallel circuit is connected in Figure 11.

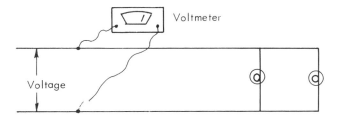

Fig. 11. Checking for Voltage on a Parallel Circuit.

Assuming no voltage drop in the circuit, the voltage will be the **same** throughout the parallel circuit. Remember, voltmeters are always connected in parallel. Voltmeters have high resistance to current flow, and if connected in series would reduce current flow to the rest of the circuit. In a series circuit, a voltmeter can be used to measure the voltage drop across each resistor or light bulb as illustrated in Figure 12.

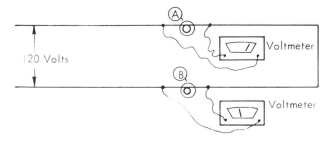

Fig. 12. Using the Voltmeter to Measure Voltage Drop Across Resistors.

The voltage drop across both bulbs should total the source voltage. For example, if the voltage drop across bulb A is 80, the expected voltage drop across B would be 40 since they must total the 120 source voltage.

$$V_t = V_1 + V_2 + V_n$$

$$V_t = 80 + 40$$

$$V_t = 120$$

The voltage in a parallel circuit is the same in all branches of the circuit, or the voltage at each branch of a parallel combination is the same as the voltage in the circuit.

Ammeters are instruments designed to measure the rate of flow of current in a circuit. Generally, two types are commonly used. They are the connected-type Figure 13 and the induction-type shown in Figure 14.

Fig. 13. Connected-Type Ammeter

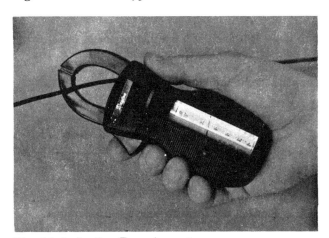

Fig. 14. Induction-Type Ammeter.

Ammeters are very sensitive instruments and easily damaged. This is especially true of the connected-type ammeter. Remember, ammeters are **always** placed in a series relationship as shown in the parallel circuit, Figure 15. If the ammeter is placed in parallel or across the 120 volts, it would be blown-out or ruined. Ammeters have very low resistance to current flow and would be damaged if connected in parallel.

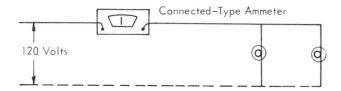

Fig. 15. Ammeter Connected in Series on a Parallel Circuit.

The induction-type ammeter is becoming more popular because it is impossible to connect it in parallel since there are no connecting lines. An induction-type ammeter is being used on a parallel circuit in Figure 16.

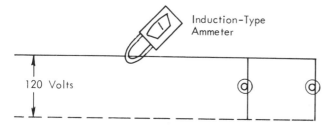

Fig. 16. Induction-Type Ammeter in Parallel Circuit.

Amperes in different parts of the parallel circuit with two or more resistors will vary. The illustration in Figure 17 demonstrates ampere flow in different parts of the circuit by the A value inside the small circles. Note, the amperages of the three individual loads can be added to equal the total amperage of the complete circuit. Therefore, amperages in a parallel circuit are **not equal throughout the circuit**, rather:

$$A_t = A_1 + A_2 + A_n$$

Current in each branch of a parallel circuit is equal to the voltage divided by the resistance of the branch. The total current in a parallel circuit is equal to the sum of the currents in the individual branches.

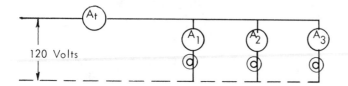

Fig. 17. A Parallel Circuit with Varying Amperages.

Ohmmeters are instruments that have their own source of power and are used to measure resistance in ohms on **disconnected** electrical resistors. In Figure 18, is an ohmmeter being used to check the resistance of a heater cone. There is little need for using the ohmmeter because ohms can be derived through mathematical use of the wattage formula and Ohm's Law formula. However, the ohmmeter is a very handy tool to check for continuity and the **worth** of heating elements, electrical motor windings, and fuses. For example, a light bulb or a fuse can be checked to determine if the bulb is blown or if the fuse element is still intact. This can be done safely without connecting these electrical units to power.

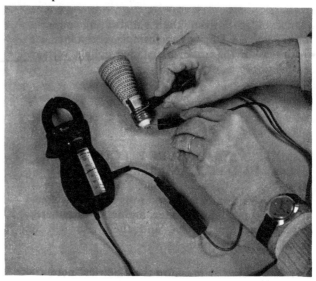

Fig. 18. An Ohmmeter Being Used to Check the Resistance of a Heater Cone.

To find the resistance of a parallel circuit the ohm value for each load must be calculated or measured. In Figure 17 three lamps are illustrated: a 40, 60 and 75 watt lamp with a 120 volt source. The individual ohms can be calculated with the **Ohm's Law formula, $R = V \div A$**. The values are 364, 240 and 192, respectively.

Calculation of the total resistance in a parallel circuit, when the values of the individual resistances are known, can be determined by one of the following equations:

Two Resistances $\quad R_t = \dfrac{R_1 \times R_2}{R_1 + R_2}$

More Than Two Resistances

$$\frac{1}{R_t} = \frac{1}{R_1} + \frac{1}{R_2} + \frac{1}{R_3} + \frac{1}{R_n}$$

The total resistance for the circuit is calculated to be

$$\frac{1}{R_t} = \frac{1}{364} + \frac{1}{240} + \frac{1}{192}$$

$$\frac{1}{R_t} = 0.00274 + 0.00417 + 0.00521$$

$$\frac{1}{R_t} = 0.01212$$

$$R_t = \frac{1}{0.01212}$$

$$R_t = 82.51 \text{ ohms}$$

The total resistance in a parallel circuit is less than any of the individual resistances.

Another common expression of Ohm's Law is that **current flowing in the circuit is directly proportional to the applied voltage,** for example:

$$A = V \div R$$

$$A = 120 \div 240$$

$$A = 0.5$$

and

$$A = 100 \div 240$$

$$A = 0.4166$$

and

$$A = 80 \div 240$$

$$A = 0.333$$

Thus, the current flowing in the circuit decreased as the voltage decreased.

Ohm's Law also states the **current flowing in a circuit is inversely proportional to the resistance,** for example:

$$A = V \div R$$

$$A = 120 \div 240$$

$$A = 0.5$$

and

$$A = 120 \div 120$$

$$A = 1$$

and

$$A = 120 \div 60$$

$$A = 2.$$

Thus, the current flowing in the circuit increased as the resistance in ohms was decreased.

Although this manual is concerned primarily with parallel circuits, the following rules must be followed when making Ohm's Law calculations with series circuits:

$$V_t = V_1 + V_2 + V_n$$

$$A_t = A_1 = A_2 = A_n$$

$$R_t = R_1 + R_2 + R_n$$

TYPES OF VOLTAGES

Voltages commonly used in North America have an electromotive force of either **120 volts, 208 volts, or 240 volts.** Industry often uses voltages of 440 volts. All of these voltages are extremely dangerous and can cause serious personal injury or death due to electrical shock. Depending upon the resistance of the human body voltages as low as 30 can be dangerous. The typical automotive or tractor battery of 6 or 12 volts does not **shock** because of its low electrical potential.

When a 120 volt current is checked by a voltmeter, the exact voltage is seldom exactly 120, but usually the measurement will be somewhere between 115 and 125. Likewise, a 240 volt circuit will not necessarily show a voltage reading of exactly 240, but may vary between 220 and 250. The same situation holds true for the 208 voltage circuits. These voltages are called high voltages because they are high enough to severely shock human beings and animals. Low voltages are in the 6 volt to 28 volt range, commonly installed in autos, tractors, low voltage control circuits, door bell circuits, and in toys such as model trains and racing cars.

A 120 volt parallel circuit is made from using one **hot** source conductor and connecting a resistor or resistors such as heaters and lamps, across it to a grounded neutral conductor, note Figure 19. This is the most common circuit arrangement used in North America. The **hot** conductor is the dangerous one because it has an electrical potential of 120 volts. It can cause serious injury; whereas, the **neutral** is going to an earthen ground and is no more dangerous than touching the earth. Note that the colors are black for the **hot** and white for the **neutral**. These are the most commonly used current-carrying conductors in application to electrical wiring. The **hot** conductors can be identified by a solid line and the **neutral** by a dashed line, note Figure 19.

Fig. 19. A 120 Volt Parallel Circuit.

The second most commonly used circuit is that which has two **hot** conductor sources, each carrying approximately 120 volts; and, whose combined voltages will be either 208 or 240 volts. Whether the voltage is 208 or 240 is dependent on the type of transformer on the service entrance. The wye type transformer provides 120/208 service and the delta type provides 120/240 service. Red and black are the colors commonly used on 208 or 240 volt circuits as illustrated in Figure 20.

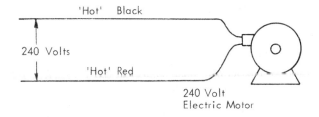

Fig. 20. A 240 Volt Parallel Circuit.

The colors used for **hot** conductors can be any color except [1] **white or** [2] **green, or bare.** The white covered conductor is a **reserved** color and must only be used as a neutral conductor that will carry current back to the earthen ground. The other **reserved** color is **green** (bare is considered the same color) and it is normally a noncurrent-carrying conductor except in cases of a circuit malfunction. Its function however, is extremely important and will be fully discussed in the Unit entitled **"Grounding for Safety."**

In summary, electricity is carried by conductors that are commonly identified by colors. A white is assumed to be a **neutral** current-carrying conductor; any color other than white, green, or bare is assumed to be a dangerous **hot** current-carrying conductor. However, this **hot** is a must because without it there is no way to provide the original pressure to force the flow of amperes through resistors to make electricity work for us.

NOTES

CLASSROOM EXERCISE I

PARALLEL AND SERIES CIRCUITS
AND
120 AND 240 VOLT CIRCUITS

1. What is the least number of current-carrying conductors required for a circuit? _____

2. The three main parts of any circuit are _____ , _____ , and _____ .

3. Circuits can be classified as either _____ or _____ circuits.

4. The symbol for a lamp is _____ .

5. In theory how many lamps or loads can be placed across the two current-carrying conductors of a parallel circuit? _____ .

6. What is a big advantage of a parallel circuit versus a series circuit? _____

7. When both series and parallel circuits are within one total circuit, this is called a _____ circuit.

8. Define:
   A. Volts _____

   B. Amperes _____

   C. Resistance _____

   D. Ohms _____

9. What letter symbol is used for each?
   A. Volts = _____   C. Resistance = _____

   B. Amps = _____   D. Ohms = (Greek Symbol) _____

10. What letter symbol is used for watts? _____

11. Define watts _____

12. Compute the amperes of an applicance that has nameplate information of 120 volts and 1320 watts.
    _____ A

13. Compute the resistance for the appliance in number 12. R = _____

14. Name the two different types of ammeters:
    A. _____   B. _____

15. Ammeters should always be connected in _____ .

16. The amperes throughout a series circuit are _____ .

17. The amperes in different parts of a parallel circuit will be?
    a. equal
    b. higher closer to source
    c. higher closer to load

18. The three commonly used voltages in North America are _____ , _____ , and _____ .

19. All voltages are considered dangerous, depending upon the resistance in the circuit, and should be respected. True-False.

20. A 'neutral' conductor leads to an _____ ground.

21. A 240 volt circuit with two current-carrying conductors will most likely be _____ and _____ in color.

22. Name the two reserved colors of electrical wiring.
    A. _____

    B. _____ or _____

23. Which reserved color is a 'neutral' current-carrying conductor? _____

24. In order to use Ohm's Law, you must have knowledge of at least _____ of the three factors.

25. Voltmeters should always be connected in _____ .

26. Complete the following problems.

    Compute ohms when:
    Amperes = 8 and volts = 120   _____

    Amperes = 15 and volts = 115   _____

    Amperes = 4 and volts = 230   _____

    Compute amperes when:
    Volts = 125 and ohms = 11   _____

    Volts = 240 and ohms = 11   _____

    Volts = 120 and ohms = 60   _____

    Compute volts when:
    Amperes = 12.5 and ohms = 9.6   _____

    Amperes = 3 and ohms = 80   _____

    Amperes = 6 and ohms = 19   _____

27. Calculate the values for the following circuit and record data in the table.

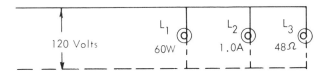

	Volts	Amperes	Ohms	Watts
Circuit	120			480
L_1		0.5		60
L_2		1.0		120
L_3			48	300

28. Calculate the values for the following circuit and record data in the table.

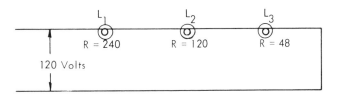

	Volts	Amperes	Ohms	Watts
Circuit				
L_1			240	
L_2			120	
L_3			48	

NOTES

LABORATORY EXERCISE I

PARALLEL AND SERIES CIRCUITS
AND
120 AND 240 VOLT CIRCUITS

DEMONSTRATIONS

Purpose: To learn differences between systems that provide 120 and 240 volts.
To learn the basic differences between series and parallel circuits.

Equipment:
1. Disconnect box: 30 ampere connected to 120/240V service
2. Receptacles: Two porcelain
3. Conductors: Black, white and red
4. Lamps: Two 40W, 120V; one 60W, 120V; one any wattage, 240V

Procedure:

1. Place two 40 watt 120 volt lamps in the 120 volt parallel circuit and operate.
   The schematic using appropriate symbols and properly labeled wire conductors appears below:

   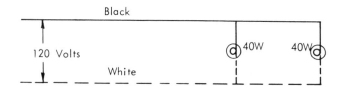

   a. What happens when only one lamp is turned out?

   _____

   _____

   b. Could other lamps be added to the parallel circuit? _____
      Where?

   _____

   _____

2. Place two 40 watt 120 volt lamps in a 120 volt series circuit and operate. The schematic is as follows:

   a. What happens when one lamp is unscrewed? _____

   _____

   _____

   b. Why? _____

   _____

   _____

3. Place one 240 volt lamp in a parallel 240 volt circuit and operate. Draw the schematic and label the conductor colors.

   a. What color of conductors were used for the 'hot' wire sources? _____ and _____ .

4. Place two 40 watt 120 volt lamps in a 240 volt parallel circuit and operate for several seconds. Draw the schematic.

   a. What happens?
   _____
   _____
   _____
   _____

5. Place two 40 watt 120 volt lamps in a 240 volt series circuit and operate. Draw the schematic.

a. What is the appearance of the lamps?

_____

_____

_____

6. Place a 40 watt 120 volt lamp and a 60 watt 120 volt lamp in a 240 volt series circuit and operate.
Draw the schematic.

a. What is the appearance of each lamp?

_____

_____

_____

_____

NOTES

LABORATORY EXERCISE II

OHM'S LAW AS RELATED TO PARALLEL CIRCUITS

DEMONSTRATIONS

Purpose: To determine Ohm's Law values in a parallel circuit.

Equipment:

1. Disconnect box: 30 ampere connected to 120/240V service
2. Receptacles: Three porcelain
3. Lamps: Two 40W, 120 volt, and one 300W, 120V
4. Heater elements: Two 600W, 120V
5. Voltmeter: AC, 0-150 scale
6. Ammeter: AC, 0-15 scale

Procedure:

1. **Volt Determination** -- Place two 40W lamps in a 120 volt parallel circuit as follows:

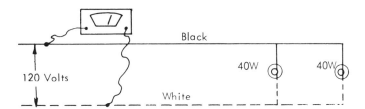

a. Use the voltmeter connected in parallel, with leads to measure the voltage at the disconnect box. _____ voltage.

b. Measure the voltage at both lamps.
_____ volts at lamp #1.

_____ volts at lamp #2

c. Define volts. _____
_____
_____

d. What general statement can be made about the voltage throughout the parallel circuit?
_____
_____
_____

e. Volts = _____ x _____ .

f. Voltmeters are to be connected in _____ .

2. **Amperes, Volts, and Ohms Determination** -- Place two 600W heater elements and a 300W lamp in a 120 volt parallel circuit as follows:

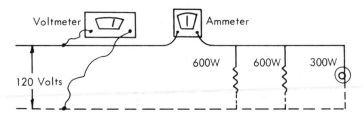

a. Make sure the ammeter is connected in series. Touch the leads of the voltmeter (in parallel) near the disconnect box.

Record: Amperes _____ Volts _____

b. Calculate resistance by using the formula:
Resistance $= \dfrac{\text{Volts}}{\text{Amperes}}$ Resistance = _____

c. Disconnect the 300W bulb and determine:
Amperes _____ Volts _____ Resistance _____

d. Disconnect one 600W heater element, leaving only one 600W heater element in the circuit and record:
Amperes _____ Volts _____ Resistance _____

e. From what you have observed in this demonstration give a general statement about amperes and load.

_____
_____
_____
_____

Give a general statement about amperes and ohm readings regarding this demonstration.

_____
_____
_____
_____

f. Ammeters are always connected in _____ . Explain.

_____
_____
_____
_____

MEASURING ELECTRICAL POWER AND COSTS

WATTS AND KILOWATTS

Electrical power is measured in **watts**. Watts can be defined as **the rate of using electrical energy or the measure of electrical power**. The consumer of electrical energy pays a bill on the basis of kilowatt-hours [kWh]. A few terms need explaining. **Kilo** means 1,000, therefore 1,000 watts equal one kilowatt. If electrical energy is used at the rate of 1,000 watts (1 Kilowatt) for one hour, then one kilowatt-hour or one kWh would have been used.

Kilowatt-hour is to electricity as **horsepower** is to an engine. Horsepower and kilowatt-hour are both determined by three factors: **force, distance, and time.** The terms used for these three factors regarding horsepower and kilowatt-hour forms of energy measurement are listed in Figure 1.

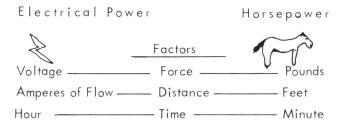

Fig. 1. Common Factors for Electrical Power and Horsepower.

METHODS OF DETERMINING POWER

There are four methods for determining power or wattage being used. They are:
- On common **resistance** devices such as electric lights and heating equipment, use the ammeter to determine flow, the voltmeter to determine pressure, and multiply the readings of both to obtain watts. This will **not** work for electric motors because they are generally induction-type electrical equipment.
- For electric motors (induction-type load) the method is similar except watts equal amperes times volts times power factor. The power factor must be calculated or assumed.
- Using a wattmeter. The wattmeter is accurate on both resistance and the induction-type equipment.
- Using a Kilowatt-hour meter. The kilowatt-hour meter is accurate on both resistance and induction-type equipment.

For each of these four methods, the equipment and meters to use, and examples will be presented in this unit.

When Watts Equal Amperes x Volts

Basically, the wattage formula or power formula is **watts equal amperes times voltage**, expressed as:

Watts = Amperes x Volts

W = A x V

A convenient wattage circle is shown in Figure 2.

Fig. 2. The Wattage Circle.

The wattage circle can be used by placing a finger over the value to be calculated when the other two values are known. For example, if wattage is 2,400 and voltage is 120, placing the finger over **A** (the value to be solved), **W** appears over **V**. Therefore,

$$A = \frac{W}{V} = \frac{2,400}{120} = 20 \text{ amperes.}$$

With the wattage formula, one can solve for the expected amperes of a light bulb, if the wattage and voltage are known. On a light bulb, information such as wattage and voltage is commonly given, but not the amperes which can be calculated as shown in Figure 3. Other electrical equipment may have only the amperes and voltages indicated. Likewise, the wattage can be calculated because two of the three factors are known.

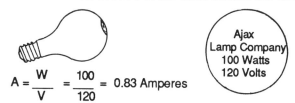

Fig. 3. Figuring Amperes of a Light Bulb.

The basic wattage formula of **W = A x V** is used only on resistance-type electrical equipment. Any electrical apparatus that basically gives off light or heat is of the resistance-type. The electric motor generally operates as an induction-type load and its formula is different. The typical hook-up to measure watts of several heater cones is presented in Figure 4. The symbol (⌇) is used for resistors that mainly give off heat, such as heater cone elements. The connected-type ammeter discussed in a previous unit is shown, but the induction-type could have been used. **The ammeter must be connected in series, if connected in parallel it would be blown out and ruined.**

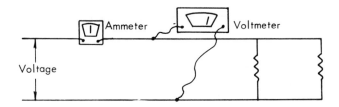

Fig. 4. Typical Hook-up for Resistance Wattage Determination Using the Ammeter and Voltmeter.

When Watts Equal Amperes Times Volts Times Power Factor

When electric motors are checked for wattage, the ammeter reading is higher than it should be and is incorrect. Therefore, an electric motor's wattage formula is:

Watts =   Amperes x Volts x Power Factor

W =   A   x   V   x   PF

The calculation of power factor will be presented later in this unit with the discussion on using the kilowatt-hour meter to determine watts. However, a 0.6 power factor is fairly accurate for fractional horsepower electric motors under one horsepower. If a 1/3 horsepower (hp) motor's nameplate information is 6.4 amperes on 115 volts, the expected estimated wattage for the motor is:

W = A x V x PF = 6.4 x 115 x 0.6 = 442 watts.

Motors that are one horsepower and over have **better** power factor; and 0.7 is fairly accurate for their power factors.

There is another method for estimating the wattage used by motors which is using the following rules of thumb:

Motors one hp and over = 1,000 watts/hp

Fractional hp motors = 1,200 watts/hp

The theoretical watts per horsepower is 746. This will help provide an idea of motor efficiency. For example, if the 1/3 hp motor uses approximately 442 watts, and by taking 1/3 of 746, the watts equal 249. However, the motor is using 442 watts; therefore, the approximate efficiency (input to output ratio) of the motor is:

Input x Efficiency = Output

$$\text{Efficiency} = \frac{\text{Output watts}}{\text{Input watts}} \times 100$$

$$\text{Efficiency} = \frac{249 \text{ watts}}{442 \text{ watts}} \times 100$$

Efficiency = 0.56 x 100

Efficiency = 56%

The electrical hook-up to determine estimated wattage of an electric motor, when a power factor will be assumed, is shown in Figure 5. Either the induction-type ammeter, or the connected-type ammeter, could be used. It is best to use the induction-type ammeter because of the high starting current of the electric motor. A connected-type ammeter (0-15 amperes range) will have its reading needle **pegged** hard to the right because typical fractional hp motors draw 20 to 30 amperes for a second or two while starting. The jaws of the induction-type ammeter can be placed around a current-carrying conductor after the motor has started, obtained its rated speed, and is drawing amperes normal to the running current of the motor.

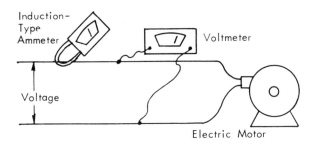

Fig. 5. Ammeter and Voltmeter Connected to Motor Circuit.

Watts Determined By Use Of Wattmeter

A **Wattmeter** is a special piece of equipment that is basically an ammeter and voltmeter built into one instrument. A typical connected wattmeter is shown in Figure 6. Note that it has four leads or points of connection. It, similar to a connected ammeter, is easy to damage if the leads representing the ammeter part of the instrument are connected parallel across the **hot** and **neutral.**

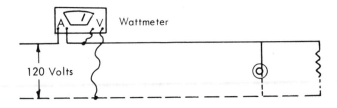

Fig. 6. Connection of a Wattmeter on a Typical 120 Volt Circuit.

Electrical equipment manufacturers provide detailed instructions on how to connect ammeters, voltmeters and wattmeters. Careful attention to these directions and instructions is a must for safety to the operator, safety in equipment operation, and long life of the instrument.

Using The Kilowatt-Hour Meter As A Wattmeter

The **Kilowatt-hour [kWh] meter** is an instrument that is found wherever electrical energy is consumed. Yet typical home and farm residents may not have a voltmeter, ammeter, or wattmeter to determine watts. The kWh meter can be used to calculate the wattage of any electrical equipment by isolating, shutting off, or otherwise disconnecting all other electrical loads except the load to be checked.

The formula used for calculation of watts, while using the kilowatt-hour meter as a wattmeter is:

Watts = rpm of meter disc x Kh [a constant, stamped on the meter] x 60

The kilowatt-hour meter can be used as a wattmeter because,

$$\text{Watts} = \frac{[\text{the rpm factor}]}{\cancel{\text{minute}}} \times \frac{[\text{the kWh factor}]}{\cancel{\text{revolution}}} \times \frac{[\text{the 60 factor}]}{\cancel{\text{hour}}}$$

For example, one could determine the wattage being used by an electric motor operating a large ventilation fan. All other electrical loads would need to be shut-off or disconnected leaving only the fan's current going through the kWh meter. Revolutions per minute [**rpm**] of the meter disc would need to be determined. It is usually best to count the disc's revolutions for at least two minutes to minimize error. Read the K_h constant stamped on the meter. Assuming the following data were collected, the motor is using 6912 watts:

Data Collected

$K_h = 6.0$

rpm = 57.6 revolutions per 3 minutes

Substituting Data into the Wattage Formula for Using the kWh Meter as a Wattmeter

$W = \text{rpm} \times K_h \times 60$

$W = \frac{57.6}{3} \times \frac{6.0}{1} \times \frac{60}{1} = 6912 \text{ watts}$

The kilowatt-hour meter is placed permanently between the power supplier's service line entering the property, and the main entrance panel or disconnect in most home and farm situations. The kilowatt-hour meter is the property of the power supplier; is sealed to prevent tampering; and, is the instrument used to meter and ultimately charge the customer for electrical power consumed. The typical kWh meter is the 120/240 volt type. This means that the meter will be connected to 240 volts but whether 120 volt or 240 volt electricity is being used, the meter does an accurate job of recording power consumed. The typcial 120/240V connection is illustrated in Figure 7.

A typical 120 volt kilowatt-hour meter connection is shown in Figure 8.

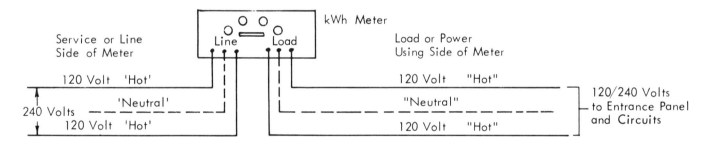

Fig. 7. A 120/240 Volt Connected Kilowatt-hour Meter.

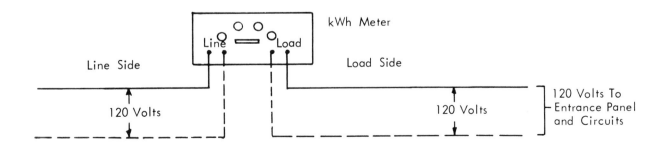

Fig. 8. A 120 Volt Connected Kilowatt-hour meter.

Note that kilowatt-hour meters have two sides--a LINE side and a LOAD side. If these sides are interchanged, the meter would run backward. This is one of the reasons the meter is sealed by the power supplier and there are laws protecting the supplier's rights against meter tampering.

COMPUTING POWER FACTOR OF A MOTOR

Earlier in this unit, it was mentioned that the amperes times voltage **does not** equal watts when an induction-type motor is involved. Rather, wattage for a motor is **watts equals amperes times volts times power factor.**

Watts [motor] = Amperes x Volts x Power Factor

W [motor] = A x V x PF

Power factor is equal to **true wattage divided by apparent wattage**. **True wattage** is obtained by using the kilowatt-hour meter; and, **apparent wattage** is obtained by using an ammeter and voltmeter.

Power Factor =

$$\frac{\text{True Wattage [determined from kWh meter]}}{\text{Apparent Wattage [amperes x volts, determined with ammeter and voltmeter]}}$$

If a motor under load is connected in the 120 volt parallel circuit as shown in Figure 9, power factor equals 0.6 because,

$$PF = \frac{rpm \times K_h \times 60}{amps \times volts}$$

$$PF = \frac{4 \times 1.8 \times 60}{6 \times 120} = \frac{432}{720} = 0.6$$

As percent of rated load changes on a motor, power factor also changes. At a no-load situation, power factor may be 0.2, but as the motor is loaded to its rated value, the power factor can be in the 0.6 to 0.8 range.

WATTS AND PAYING FOR THEM

Since watts is power and the cost of using power is a very real situation, an understanding of how electricity is purchased is important. An example will help explain an electric bill. If a room is heated with an electric baseboard heater that is connected to 240 volts, drawing 10 amperes and operating 5 hours per day, for one month (30 days), with an electrical power charge of 3c per kilowatt-hour, the consumer will pay $10.80 because:

W = A x V

W = 10 amperes x 240 volts = 2,400 watts

$$\frac{2,400}{1,000} = 2.4 \text{ kilowatts}$$

2.4 kilowatts x 5 hours per day = 12 kWh per day

12 kWh/day x 30 days = 360 kWh

360 kWh at 3c = $10.80.

The electric bill is normally paid monthly by the consumer. The kilowatt-hour meter is the instrument that accurately records the consumption of electricity used, and can be compared to the odometer of an automobile that accurately records the miles traveled. Two types of kilowatt-hour meters, note Figure 10, are the **cyclometer-type** and **pointer-type**.

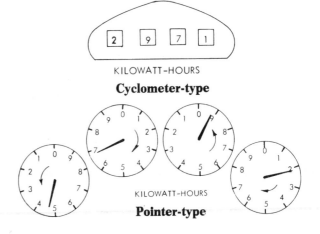

Fig. 10. Types of Kilowatt-hour Meters.

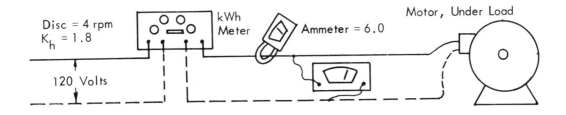

Fig. 9. Determining Power Factor of a Motor.

The pointer-type is more difficult to read because every other clock goes in the opposite direction. Each pointer must be studied to determine which numbers are read. The pointer between 6 and 7 is read as 6. The pointer-type meter in Figure 10 is reading 4692.

Some power suppliers have service employees who read kilowatt-hour meters on a once-a-month basis for consumer billing purposes. However, in some parts of the country, consumers are asked to read their own meters, record the information on a card, calculate their own power costs, and pay their bill according to the rate schedule. Rate schedules commonly used in the United States are on a graduated rate basis--which means that the first few kilowatts cost more than the kilowatts used later in the month. One power supplier utilizes the following rate schedule for computing electrical bills.

Quantity	¢/kWh	$
1st 30 (minimum)	--	4.00
Next 70	8	5.60
Next 100	4.5	4.50
Next 400	2.5	10.00
All over 600	2.3	

Fig. 11. A Graduated Rate Charge Schedule.

A consumer using less than 30 kilowatt-hours, see Figure 11, would still pay the minimum $4.00 charge. This minimum charge is justified because power companies have very high fixed generation and power transmission costs. Yet they provide service to both small and large power consumers. The major variable costs to power companies are coal, natural gas, and nuclear reactor fuels.

If a consumer purchasing energy from the power company in Figure 11, used 643 kilowatt-hours during a month, the bill would be $25.09. Table 1 explains why. However, if another consumer used only 244 kilowatt-hours, the consumer's bill would be $15.20, because the first 200 kilowatts cost ($4.00 + $5.60 + $4.50), and the 44 kilowatts over 200 cost, 2.5 cents each, which is $1.10 for a grand total of $15.20.

Table 1. Example of Consumer Using 643 Kilowatt-hours.

Quantity	¢/kWh	$
1st 30 (minimum)	--	4.00
Next 70	8.0	5.60
Next 100	4.5	4.50
Next 400	2.5	10.00
Next 43	2.3	0.99
643 kWh		$25.09

NOTES

CLASSROOM EXERCISE II

MEASURING ELETRIC POWER AND COSTS

1. Define watts: _____

    _____

    _____

2. There are _____ watts in a kilowatt.

3. If 2000 watts are used for 3 hours, how many watt-hours would be used? _____ How many kilowatt-hours (kWh) would be used? _____

4. Electrical equipment that produce light or heat are called _____ loads.

5. List the wattage formula for a resistance load.

6. The wattage formula for a motor, induction type load is stated as:

7. List the four different pieces of equipment used for determining electrical power.
    A. _____ B. _____
    C. _____ D. _____

8. Calculate the amperage of an electric heater designed to deliver 1800 watts when connected to 240 volts.

    _____ A

9. The power factor on smaller motors is generally (higher or lower) than those of larger motors.

10. Theoretical watts per horsepower is: _____ .

11. Calculate the efficiency of a 1/2 horsepower motor that is using 471 watts.

12. Motors can be estimated to use, according to a rule of thumb, about _____ watts per horsepower on fractional hp motors, and _____ watts on motors one hp and over.

13. Calculate the wattage of a motor that draws 6.2 amps, on 120 volts with an assumed power factor of 0.7.

14. A wattmeter has connections for _____ and _____ .

15. Compute the wattage of an appliance, when the rpm of the meter disc on a kilowatt-hour meter is 12 and Kh value is 3.6

16. Read the two pointer-type meters. If the one on the left represents the first of the month and the one on the right represents the end of the month, calculate the kilowatt-hours used. _____

    First of the Month   End of the Month

17. Compute the monthly electrical cost for the kilowatt-hours used in Question 16, using the monthly rate schedule in Figure 11.

    $ _____

18. Compute the monthly (30 day) electrical bill using the rate schedule in Figure 11 when the following load is used:
    1. Lighting - 1200W for six hours per day.
    2. Appliance - 2048W for one hour per day.
    3. Two electric motors, each drawing six amperes on a 120 volt circuit and each running an average of one hour per day. Assume 0.8 power factor.
    4. An electric water heater drawing 8 amps on a 240 volt circuit and operating from 1 to 6 p.m.

NOTES

LABORATORY EXERCISE III

MEASURING ELECTRIC POWER AND COSTS

DEMONSTRATIONS

Purpose: To learn how to use meters in determining wattage, watt-hours, and kilowatt-hours, and to calculate power costs.

Equipment:

1. Disconnect box: 30 amperes connected to 120/240V service
2. Receptacles: Three porcelain
3. Lamp: One 40W, 120V
4. Heater Elements: Two 600W, 120V
5. Motor: One 1/3 hp split-phase operating on 120V
6. Ammeter: AC, 0-15 scale
7. Voltmeter: AC 0-150 scale
8. Wattmeter: AC, 0-1500 scale (optional)
9. Kilowatt-hour meter: 120V

Procedure:

1. Place two 600W heater elements and a 40W lamp in a 120V parallel circuit as shown:

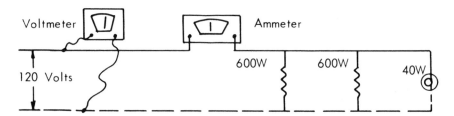

    a. Determine the voltage and amperes by use of the voltmeter and ammeter. Make sure the ammeter is connected in series. Use the voltmeter with leads to measure voltage by making contact near the source. Reading on voltmeter = _____ . Reading on ammeter = _____ .

    b. Watts equal _____ x _____

    c. Calculate the wattage being used in the experiment.
Watts = _____ x _____ = _____

    d. Define watts: _____

_____

_____

2. Using the same load as in Number 1, measure the wattage with a wattmeter. Make sure the wattmeter is connected as shown.

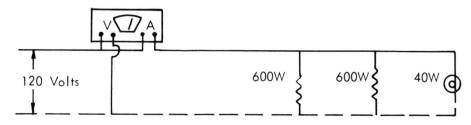

    a. Record reading. _____ watts

b. Did this vary from the previous wattage calculated when the ammeter and voltmeter were used?_____ . If so, explain.

_____

_____

_____

_____

3. Using the same load as in Number 1 and 2, measure the wattage with a kilowatt-hour meter. Make sure it is connected as shown.

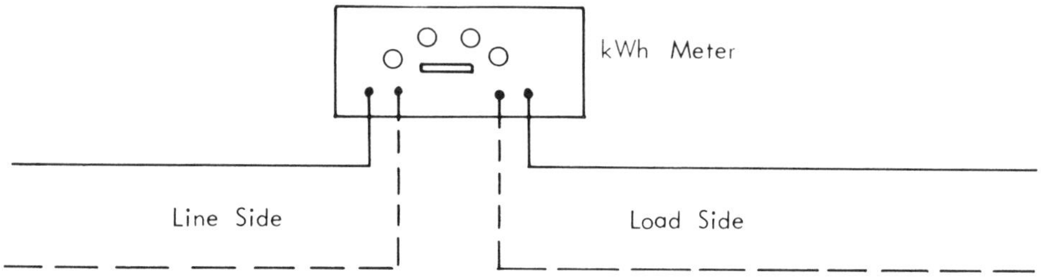

a. Use the formula:
Watts= rpm of meter disc x K_h x 60, operated for several minutes to determine rpm of meter disc and then determine the watts being used. _____ watts.

b. Assuming this load is operated for six hours, how many watt-hours would be used?_____ How many kilowatt-hours?_____ If kWh cost 3c, what would be the cost of operating this load for six hours? _____

4. Using a Prony brake or another method of loading, operate a 1/3 hp split-phase motor with a load of 7 amperes on 120 volts. Make sure the load is maintained as nearly constant as possible for one minute while the rotation of the kWh meter disc is counted. Show how the circuit should be set-up with the ammeter connected in series, and the voltmeter in parallel; and kWh meter between source and load.

a. Calculate the true wattage by using the kWh meter.

_____ rpm x _____ K_h x 60 = _____

b. Calculate the apparent wattage by multiplying amps x volts obtained from the ammeter and voltmeter.

_____ Amperes x _____ Volts = _____

c. Use the following formula to compute power factor:

$$PF = \frac{\text{Watts (true wattage, determine from kWh meter)}}{\text{Amperes x Volts (apparent wattage)}}$$

$$PF = \frac{\rule{3cm}{0.4pt}}{\rule{1cm}{0.4pt} \text{ x } \rule{1cm}{0.4pt}} = \rule{3cm}{0.4pt}$$

5. If an electric water heater had its nameplate information missing, how might the wattage be determined?

NOTES

SWITCHING FOR ELECTRICAL CIRCUITS

The **switching** of electrical conductors to control lamps and other loads is of prime importance in electrical wiring. Switching and its application to practical wiring can be simple, yet in some cases, such as using a combination of switches, it can be complex.

Switches are basically directional devices, because they can stop the flow of current in a conductor, or they can direct electrical flow along another path. Before discussing specific applications of switches it is necessary to first become familiar with some basic terms related to types and kinds of switches. Illustrated in Figure 1 are terms basic to most switches or control devices and schematics of commonly used switches.

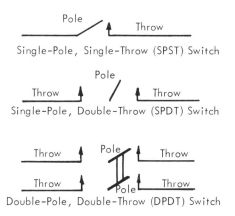

Fig. 1. Common Switch Terms and Switches.

SINGLE-POLE SWITCHING

The simplest and most commonly used switch is the **single-pole, single-throw switch** that either permits current to flow or stops the flow. The single-pole, single-throw (**SPST**) switch commonly used to control 120 volt circuits, is often called a single-pole switch and is symbolized as either **S** or **S₁** as shown in Figure 2. The switch illustrated is in the **open** position. Open, in this application can also mean **off**. In fact, single-pole switches usually have the letters **on** and **off** imprinted on the tumbler lever.

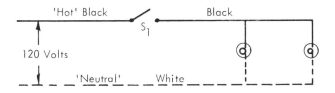

Fig. 2. Using a Single-pole Switch to Control Two Lamps.

Illustrated in Figure 3 is a parallel circuit with the switch **closed** and the lamps **on**. Note that the switch would intercept or break current flow on the black conductor, which is the **hot** conductor and carrying the 120 volts. Several rules pertinent to practical wiring of switches in a circuit can be stated:

- The switch must be in the **hot** or feed conductor.
- The switch is in a **series** relationship within the conductor.
- The white **neutral** is never switched.

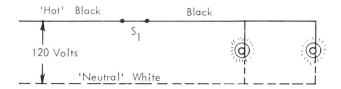

Fig. 3. A Single-pole Switch in the Closed Position.

The colors of the terminals on switches are always **brass** or **bronze**. This color coding directly implies that switches are placed in series as they intercept a **hot** feeder conductor.

The positioning of a switch within a circuit may be:

Source ⟶ Switch ⟶ Fixture

as illustrated in Figures 2 and 3. However, another very common circuit arrangement is:

Source ⟶ Fixture ⟶ Switch

This arrangement is known as a **switch-loop circuit**. From a schematic standpoint, it should be connected as illustrated in Figure 4.

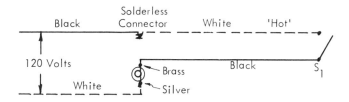

Fig. 4. A Typical Switch-loop Wiring Schematic.

In practical wiring, it is a must that the white **neutral** goes unswitched from source to the silver terminal of a fixture such as a lamp. In addition a black or some color denoting **hot** must go to the brass terminal of the fixture. Most commonly used electrical cables have a black conductor and white conductor. When this cable is used in the switch-loop part of the circuit, as shown in Figure 4 (the circuit part from the lamp area to S₁ area) a white conductor will be joined to the black source conductor. The joining will usually be done with a solderless connector or wirenut. This white conductor is then considered to be black because it is placed between the black source and the switch, as discussed in Section 200-7 of the National Electrical Code (NEC) provisions. One rule of thumb applicable to this situation is that the **neutral** is always **white**, but the **white** is not always **neutral.**

THREE-WAY SWITCHING

A **three-way switch** is also a directional device that causes the path of current to be directed along one path, or along another path. A schematic to help explain is presented in Figure 5, 6 and 7.

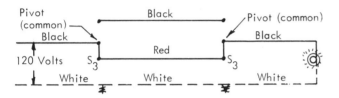

Fig. 5. Two, Three-way Switches Positioned to Light the Lamp.

In Figure 5 note that each three-way switch is **flipped** so that a pole connects from the common terminal **pivot** downward. The current will flow in the red conductor to complete the circuit and light the lamp. If either switch is **flipped** (assume the S3 on the left), this will open the circuit and the lamp will go out as illustrated in Figure 6.

Fig. 6. Three-way Switch with Lamp Off.

If the switch on the right in Figure 6 is **flipped**, the lamp will light again as illustrated in Figure 7.

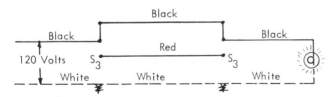

Fig. 7. Three-way Switches with Lamp On.

In summary, three-way switches are **single-pole, double-throw (SPDT)** switches that are specially wired or connected for an intended job or purpose. The purpose of three-way switching is to control a load or several loads by using **two** three-way switches positioned in different locations of a circuit.

Figures 5, 6 and 7 illustrated a circuit arrangement of:

Source ⟶ Switches ⟶ Fixture

However, a switch-loop arrangement, whereby the circuit is:

Source ⟶ Fixture ⟶ Switches

can also be wired as illustrated in Figure 8. The connected lamp is on because the circuit is **closed.** If either three-way switch is flipped, the circuit would be **open** and the lamp would go out. Note the use on conductor colors and terminal colors throughout the circuit. This is the best way for handling these colors in meeting NEC minimal requirements. Also note that a three conductor cable is used between the two S3 switches, as in Figures 7 and 8, yet the rest of the circuit requires only two current-carrying conductor cables.

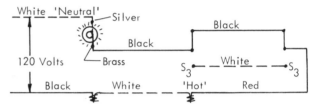

Fig. 8. A Switch-loop Application of Three-way Switching.

In addition to the basic rules for switches given earlier a rule that must be followed in wiring three-way switches in a circuit states: The common terminal **(pivot)**, generally a darker color than the other two terminals, must be connected to either **(1) the black conductor leading directly from the source (feeder), or (2) the black conductor leading directly to the brass terminal or hot of the lamp.** Study Figures 5-8 to note use of this rule.

FOUR-WAY SWITCHING

Four-way switches are used in applications whereby a load is to be controlled from three or more locations. Two three-way switches must be used whenever one or more four-way switches are used as in Figure 9.

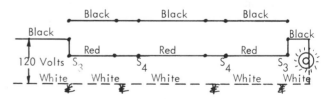

Fig. 9. Controlling a Lamp From Four Locations Using Two, Three-Way Switches and Two, Four-Way Switches.

A four-way switch (S4) is basically a specially wired **double-pole, double-throw (DPDT) switch** designed to pass current directly through (Figure 9) or to criss-cross the flow as illustrated in Figure 10.

Study Figure 9 and trace the current flow from the black source through all four switches, through the lamp and back to source white. Now trace the current flow through Figure 10, because the first S4 has been **flipped.** Note that the lamp goes out. **Flipping** any switch (whether S3 or S4) will cause the lamp to either light or go off.

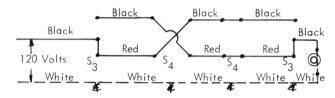

Fig. 10. Tracing the Circuit to Show That It Is Open.

Note that in Figures 9 and 10 a **three conductor cable** is used between all three-way and four-switches, yet a **two conductor cable** is used between the source and the first three-way and between the last three-way switch and the lamp fixture. Solderless connectors (⌇) are shown in these schematics because in practical wiring they are used to make cable splices. All splices must be made within the confines of an electrical box. Also, all devices such as electrical switches must be mounted to electrical boxes. More information on cables, splices, and mounting of electrical devices in boxes will be covered in a later unit.

NOTES

CLASSROOM EXERCISE III

SWITCHING FOR ELECTRICAL CIRCUITS

1. Switches are considered _____ devices.

2. The simpliest of all switches is the _____ .

3. The symbol used for a single-pole switch is _____ or _____ .

4. When a switch is open, the lamp is (on-off). When the switch is closed, the lamp is (on-off).

5. Draw a switch-loop with two parallel connected lamps between 120 volt source and a single-pole switch.

6. How many current-carrying conductors are in a circuit with two, three-way switches
    a. Between source and the first three-way switch? _____
    b. Between the two, three-way switches? _____
    c. Between the second three-way switch and the lamp fixture? _____

7. In a multiple switching system, the four way switches are always located between the _____ switches.

8. Draw a schematic showing:
    Source ——▶ Switches ——▶ Lamps
   using two, 3-way switches, three, 4-way switches and two lamps parallel connected. Show the first and third 4-way switches as 'criss-crossed'. Sketch with the lamp's light 'on'.

LABORATORY EXERCISE IV

SWITCHING FOR ELECTRICAL CIRCUITS

DEMONSTRATIONS

Purpose: To learn how single-pole, single-throw (simple toggle switches), single-pole double-throw (SPDT or 3-way switches), and double-pole double-throw (DPDT or 4-way switches) operate.

Equipment:
1. Disconnect box: 30 amperes connected to 120/240V service
2. Switch: One single-pole, single-throw
3. Switches: Two, 3-way
4. Switches: Two, 4-way
5. Lamps: Two, 40W, 120V

Procedure:

Set up a 120 volt parallel circuit for all demonstrations in this lesson.

1. Controlling a lamp with a single-pole single-throw switch. The symbol for this switch is ╱ • (open) or ━━━ (closed).
   Draw the schematic showing the circuit as open and label conductor colors.

   Should the neutral wire be switched? _____

2. Controlling two lamps with a single-pole switch. Draw the schematic showing a 'closed' circuit and label conductor colors.

   Could additional lamps or receptacles be added and controlled by the single-pole switch? ____
   Where?

   _____
   _____
   _____
   _____

3. Controlling a lamp with 3-way switches. The symbol for a 3-way switch is ⁀ or ⁀

   Draw the schematic showing the circuit as 'open' and label the conductor colors.

   a. Add a duplex receptacle not controlled by either switch. Draw the schematic with the circuit 'closed' and label conductor colors.

   b. How many conductors should a cable contain when connected between two, 3-way switches? _____ What colors could they be? _____ , _____ , and _____ .

4. Controlling a lamp with two, 3-way switches and one 4-way switch. The symbol for a 4-way switch is ⁀ or ✕ .

   Draw the schematic showing a 'closed' circuit and label conductor colors.

   a. Add another switch. What type and where in the circuit should it be added?
   _____
   _____
   _____

   b. How many 3-way and 4-way switches would be used to control a lamp from six locations?
   _____ 3-ways and _____ 4-ways
   From five locations? _____ 3-ways and _____ 4-ways

GROUNDING FOR SAFETY

THE GROUNDING SYSTEM

In previous Units, circuits have been discussed. However, the conductors used in these circuits have been presented with only current-carrying conductors, and nothing has been mentioned regarding the so-called **third-wire** conductor or the **grounding** conductor. The grounding conductor is an additional conductor that is not needed for the proper functioning of the basic circuit, but is a separate noncurrent-carrying conductor system mainly provided for two reasons:

- To direct current flow to an earthen ground to protect **equipment**, and
- To direct current flow to an earthen ground to protect **people** and **animals**.

Illustrated in Figure 1 is a basic circuit with the **third-wire (grounding)** system in place.

A grounding-type duplex receptacle has two silver screws on one side (**the white neutral attachment side**); two brass screws on the other side (**the 'hot' attachment side**); and, a green grounding screw which is in electrical continuity to the two small holes on the face of the receptacle. Therefore, when an appliance, such as an electric drill, is plugged into the receptacle and the black or **hot** conductor is shorted out to the frame the current will flow from the source, through the grounding conductor (**bare or green**) and **blow** the fuse or trip the ciruit breaker as shown in Figure 2, unless the individual's body has less ohms resistance than the conductor.

However, the **short** potential in terms of amperes flow is sometimes not enough to blow the fuse (**less than 15 amps, for example**) and then the current will flow back to earthen ground through the bare or green conductor. Extra electricity would be used, but the operator of the drill would not get a shock or learn that the drill is shorting out, or know that the drill is dangerous and needs repair or discarding. But, **if** the bare or green conductor is **faulted** or broken somewhere in the system, as in Figure 3, the electrical potential of the shock could then be carried off to ground through the operator and to the earthen ground.

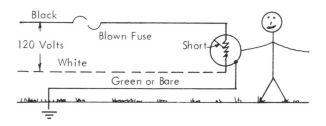

Fig. 2. Excessive Amperage Blowing a Fuse Due to a Motor Short.

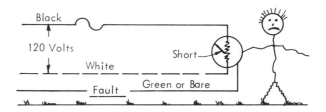

Fig. 3. A Motor Short with a Fault in the Grounding System.

This faulted situation could cause serious electrical shock, skin burns or death. The human body, even if it could carry enough amperes to **blow** the fuse, could experience heart stoppage at around **15 to 50 milliamperes or 0.015 to 0.050 amperes**, which is far below common fuse sizes. Maintaining the grounding system with the **third-wire** is for safety, and not to do so is foolish. In fact, the National Electrical Code (NEC) has for over 20 years made the grounding system for most wiring applications mandatory for industry and the general public of cities, towns, county and other agencies who officially through law accept the NEC provisions as minimal standards for safety. Yet fool-hardy people still cut off the **third-prong** of power cord caps which is an open invitation to electrical shock.

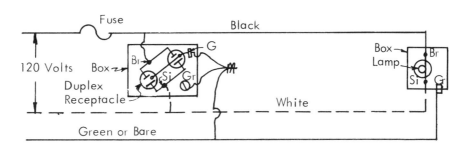

Fig. 1. Grounding with the Bare or Green Conductor on a Typical 120 Volt Circuit.

Some manufacturers are producing power tools which do not have the **third-prong** on the power cord, under the theme of **double-insulated**. These manufacturers claim that electrical parts such as switches, brush holders and motor windings can not logically in normal breakdowns short through to outside tool cases and cause electrical shock. However, this is very debatable. In summary, these statements can be made regarding grounding or the use of the **third-wire** conductor:

- When grounding is intact, big shorts (load greater than fuse capacity) blow fuses (SAFE).
- When grounding is intact, small shorts (load less than fuse capacity) may or may not blow fuses (SAFE).
- When grounding is **not** intact, all shorts have the potential of killing or causing serious electrical shock to humans and animals (UNSAFE).

THE GROUNDING SYSTEM APPLIED TO 240 VOLT APPLICATIONS

The typical 240 volt circuit, illustrated in Figure 4 demonstrates grounding. Used in this manner, it is usually referred as equipment grounding. Note that two fuses, or current overload protection devices, are in both **hot** lines. No **neutral** current carrying conductor is used or required.

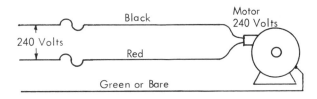

Fig. 4. Equipment Grounding a 240 Volt Connected Motor.

Certain appliances, such as clothes dryers, that are connected to 240 volt lines must have a current carrying **neutral** to make them operative. Even though many appliance manufacturers join the **neutral** to the appliances inside the case or appliance frame, the additional equipment grounding diagramed in Figure 5 is a good idea and affords extra protection to equipment and humans. In certain electrical applications this arrangement is required by law, as with some state laws regulating industry, public schools and other agencies.

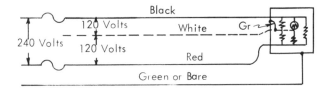

Fig. 5. A 120/240 Volt Appliance Wired with a Grounding System.

The farmer needs to constantly guard against the possibility of electrical shock to animals. Grounding does this, but the threat of a grounding fault is always present. Therefore, a ground rod driven into the earth near the placement of the appliance is a good idea. One example of this application is with the electric water heater drinking fountain illustrated in Figure 6.

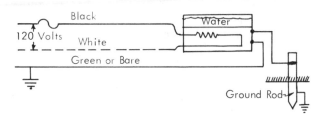

Fig. 6. Using the Driven Ground Rod as Backup to Grounding.

The **ground rod** acts as a backup system to the grounding conductor. It can, in the event of a grounding fault save cattle, hogs and other livestock species which are highly susceptible to death due to electrical shock. The ground rod is usually made of steel with a thin outer covering of a copper alloy. Special grounding clamps are used to join the conductor to the rod, which is usually eight feet long and one-half inch in diameter. The ground rod to equipment frame should have an ohmmeter reading of 25 or less ohms to be considered an adequate ground.

THE GROUND-FAULT CIRCUIT INTERRUPTER

Recently a new device called a **ground-fault circuit interrupter [GFCI]** has come into prominence. It was first mandated by NEC in 1971 for use on residential 15 and 20 ampere outdoor outlets, with portable swimming pools, and with outlets located within 10 to 15 feet of swimming pools. In 1975 requirements were broadened to include bathroom receptacles. Several styles of ground-fault circuit interrupters are available on the market, namely:

- The circuit breaker style that mounts in the service entrance panel, protecting either a branch circuit or the total service entrance panel.
- The duplex receptacle style that mounts in a common electrical box.
- The GFCI style that plugs into a duplex receptacle and protects users of electrical equipment plugged into the GFCI receptacle.
- The portable GFCI, with power cord attached, designed to serve as an extension cord for on-the-job equipment.

The four styles are illustrated in Figure 7.

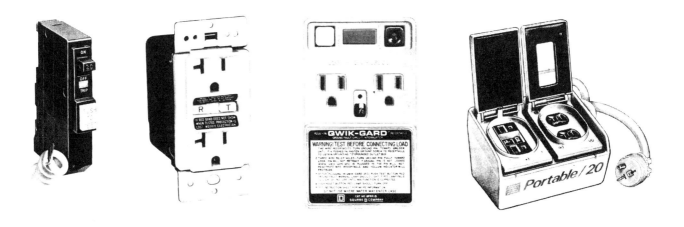

Fig. 7. Different Styles of GFCI's.

Basically, the GFCI consists of a **doughnut** differential sensing transformer that detects any current flowing to ground. Solid state components amplify this current sufficiently to actuate a mechanism that opens or trips the circuit on the **hot** conductor line. Actually the differential transformer senses a current imbalance or a ground-fault, and some of the current **going out** is not **coming back** to the source. In a 120 volt circuit this would mean that some of the current going out in the black is not returning through the white. Study Figure 8 for further understanding of the operation of the GFCI.

A normal circuit breaker or fuse does not have this differential sensing transformer and therefore can not measure the **imbalance**. The type of fault indicated by this imbalance is extremely dangerous to human life because it puts an electrical potential on metal parts of equipment that may be contacted by people. The GFCI detects this imbalance at levels well below that of human tolerance. It does so with incredible speed. The **electrocution level** for both men and women is **380 milliamperes (0.380 of an ampere)** for a duration of **1/10 of a second.** The GFCI devices in use today sense and open the circuit within **25 milliseconds [1/40 of a second]** when the current of the ground fault reaches **5 milliamperes [5/1000 of an ampere].**

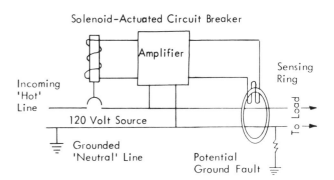

Fig. 8. Operation of the GFCI in an Electrical Circuit.

In summary, the GFCI has two purposes which are **to save lives** and **reduce possible injury** from electrical shock. The only disadvantage is that they increase the cost for a specific electrical circuit.

CLASSROOM EXERCISE IV

GROUNDING FOR SAFETY

1. The grounding conductor is an _____ conductor and is not needed for functioning of the basic circuit.

2. Is the grounding conductor normally a current carrying or a noncurrent carrying conductor? _____

3. When a 'short' occurs in a machine, the grounding conductor should carry the voltage from source to _____, placing an excessive load on the fuse causing it to _____

4. When a 'short' occurs in a machine, and there is no grounding conductor, what could happen to a person touching the machine?

5. Would the human body be likely to carry enough amperage to 'blow' a common fuse? _____

6. On a typical 120 volt circuit, the fuse is placed on the (black, white or green) conductor?

7. On a typical 120/240 volt circuit the _____ and _____ colored conductors would be fused. Should the bare or green be fused? _____ Should the white neutral be fused? _____

8. Why is using a ground rod near and connected to an appliance, such as a livestock drinking fountain a good idea?
   _____
   _____
   _____
   _____

9. GFCI stands for _____

10. Name two styles of GFCI's.
    A. _____
    B. _____

11. The GFCI works on the principle that it will trip the circuit when its differential transformer senses a current _____ or a _____.

12. The purpose of the GFCI, from a human standpoint, is to _____ and _____.

NOTES

LABORATORY EXERCISE V

GROUNDING FOR SAFETY DEMONSTRATIONS

Purpose: To learn the methods of polarization in 120 volt and 240 volt circuits.

To learn differences between grounded and grounding.

Equipment:
1. Disconnect box: 30 amperes connected to 120/240V service
2. Parallel grounded duplex receptacle with a box that has provisions for grounding
3. Lamp receptacle with a box that has provisions for grounding
4. Single-pole switch with a box that has provisions for grounding
5. Conductors: Black, white, red, and green or bare conductors to make proper connections
6. Lamp: Any wattage, 120V
7. Dual voltage motor wired for 240V service

Procedure:

Set up a 120 volt circuit with one receptacle and lamp, properly using the grounding system. Draw the schematic and label conductor colors.

a. What conductor color was used for:
   hot?_____; neutral or grounded?_____; grounding?_____

b. What color was the receptacle terminal used to connect to:
   black conductor?_____; white conductor?_____; bare or green conductor?_____

Where else was the bare or green conductor connected in addition to the duplex receptacle?_____

What connecting device was used to complete this connection?_____

2. Set up a 120 volt circuit with a single-pole, single throw switch controlling a lamp. Place the switch between source and lamp. Draw the schematic and label conductor colors. Be sure to make provisions for proper grounding.

a. What color is the two terminals on the switch?_____

What colors are the terminals on the lamp base?_____

and_____

b. Where should the bare or green conductor be connected?
_____ and _____

c. In the above installation the cable should contain how many conductors? _____ Explain.
_____
_____

3. Set up a 120 volt circuit with a single-pole, single-throw switch controlling a lamp that is between the switch and the source of power. Draw the schematic and label conductor colors. Show proper grounding.

   a. Does this circuit require that the black and white to be connected together? _____ Explain.
   _____
   _____
   _____

4. Set up a 240 volt circuit and operate an electric motor. Draw the schematic and label conductor colors. Be sure to consider grounding provisions.

   a. List the colors of conductors supplying the hot source.
   _____ and _____

   b. What is the color of the grounding wire? _____ or _____

   c. Where in the system should the grounding wire be connected?
   _____

PROVIDING FOR ADEQUATE WIRING AND OVERLOAD CURRENT PROTECTION

WIRE SIZES, CONDUCTORS AND INSULATORS

Electric power flows through **wire conductors.** It flows more easily over some materials than others. Conductors are defined as **materials having low resistance to electrical current flow.** Materials such as **gold, silver, copper, aluminum, mercury,** and other metals such as **steel and iron** are excellent conductors. The opposite to conductors is **insulators.** Excellent insulators are **plastic, glass, porcelain, rubber, paper and wood.** Insulators retard or inhibit the free flow of electrical current. Thus, they are considered to have high resistance to the flow of electrical current. Electrical conductors commonly have a wire core, such as copper, surrounded by an insulation, such as plastic, to help insulate the conductor from **shorting** into electrical boxes and other metallic parts of a wiring system. Copper is the best practical material for conductor use. All reference made in this unit is to copper conductors.

Copper wire sizes are indicated by **number.** Number 12 and 14 are the most common sizes used for residential wiring. A number 14 is larger than a number 16; and, a number 8 is smaller than a number 6, as shown in Figure 1. Numbers 16 and 18 are used mostly with flexible (stranded wire) power cords. Smaller sizes are used in the manufacture of electrical appliances such as motors.

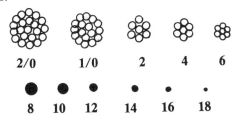

Fig. 1. Actual Size of Copper Conductors without Insulation.

AMPACITY

Wire of the correct size must be used in circuits for two reasons: **ampacity** and **voltage drop.** Ampacity is the **safe carrying capacity of a wire, in amperes.** Due to resistance of current flowing through a conductor a certain amount of heat is produced. This heat is wasted; therefore, to avoid wasted power, wire size should be large enough to keep waste to a reasonable and practical level, yet small enough to carry the load. When amperages are much too large for the size of the conductor used, the wire can become so hot that fire is a possible hazard. The National Electrical Code (NEC) is not concerned with wasted power, as such, but is concerned with safety. Therefore, it sets ampacity, or maximum amperage that various sizes and types of wires are allowed to carry.

VOLTAGE DROP

If forcing too much amperage through a wire caused only a certain amount of wasted power, it might be looked upon as a mere nuisance. However, it also causes **voltage drop.** Actual voltage is lost in the wire so that the voltage across the circuit is lower at the load end of the circuit than at the circuit source. An example of this is illustrated in Figure 2. The difference in voltage of 120 at source and 114 at the motor is called voltage drop. Voltage drop is wasted power, and another important consideration is the inefficient operation of electrical equipment on a lower than normal voltage. At 90 percent of rated voltage, a motor produces only 81 percent of its normal power, and a lamp produces only 70 percent of its light. Percent voltage drop is determined by:

Percent Voltage Drop =

$$\frac{\text{Source Voltage Minus Load Voltage}}{\text{Source Voltage}} \times 100$$

Voltage drop is a combination of three factors:
- Load
- Length of Run
- Size of Wire

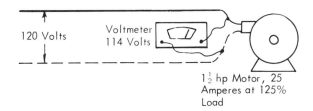

Fig. 2. An Example of Voltage Drop from Source to Load.

A number 6 conductor is required for the 25 ampere motor illustrated in Figure 2, to keep the voltage drop to a reasonable level (3 percent), when the distance is 125 feet. This can be determined from data in Table 3 found in the Appendix. However, to limit the voltage drop to a 2 percent, a number 4 wire would be required. If a 4 percent voltage drop was acceptable, number 8 wire conductors could be used. Obviously, the voltage drop for the circuit in Figure 2 would be inadequate because 120 volts minus 114 volts divided by 120 volts times 100 equals a 5 percent voltage drop which is unacceptable; and, the conclusion can be made that the conductor wires selected were too small. For residential and farm wiring the system should be designed for a 2 to 4 percent voltage drop.

Most wire size problems can be solved by data provided in Tables 1 through 5 of the Appendix. For example, compute the size of conductors needed for service lines to a water heater that is 75 feet from source, and draws 1200 watts on 120 volts. First, the expected current would need to be determined as:

$$A = \frac{W}{V} = \frac{1200}{120} = 10 \text{ amperes}$$

Whether the installation is underground burial or overhead in air becomes a factor. If the 3 percent voltage drop is accepted, then number 12 wire conductors would be used for the underground burial, and a type **UF [underground feeder]** would be installed. However, if this run was done with a single span of 75 feet (pole to pole), a number 8 conductor would be required. If a pole is placed in the center of the span, a number 10 could be used. The physical strength of the conductor is a factor in solving this problem relative to overhead conductors. One advantage of underground burial of conductors is that on many light load applications where medium and long runs are involved, wire size needed is often smaller reducing the cost of the conductor materials. However, the cost of trenching is usually a trade-off factor with the reduced conductor cost. Other advantages of underground circuitry are that conductors are out of the way and safer regarding movement of equipment and accidental contact by people.

There is always a wire size advantage in regard to equipment using the same wattage or horsepower in connecting the equipment on 240 volts rather than 120 volts. If installing a 1/3 horsepower motor with conductors underground or in a conduit raceway and allowing a 2 percent voltage drop, a number 12 wire would be required for a 100 foot run when connected to 240 volts; whereas, a number 10 wire would be needed when connected to 120 volts.

PROVIDING PROPER OVERLOAD CURRENT PROTECTION

Providing proper feeder wire sizes or conductors to electrical equipment has been discussed in this unit. Closely associated to this requirement is proper fusing or overload current protection. Consider for example, the isolated motor that is unattended while operating. If this motor were accidentally overloaded and were to draw excessive amperages well above rated amperages, it would try to continue running until it **burned** out. However, if properly fused, the fuse would detect this overload and **blow** or open the circuit. The current flow to the motor would then be stopped, preventing damage to the motor.

The term **fuse** is a very general term that can apply to any type of overload current protector, such as Edison base fuses (both standard and time-delay), circuit breakers, circuit breaker-style ground-fault circuit interrupters (GFCI), cartridge-style fuses, and overcurrent safety devices built into the source leads of the motor.

OVERLOAD CURRENT PROTECTION FOR MOTORS

Generally motors should have **fusing** devices that are of a **time-delay** action type to allow for the high starting amperages before the motor reaches its operating rpm and its running amperage. For example, a typical 1/3 horsepower split-phase motor may draw 20 to 30 amperes for about 2 to 4 seconds when starting; and, then draw 5 to 6 amperes at its rated full load when running at full or rated rpm. Therefore, proper fusing of most motors is at 100 to 125 percent of full load. With motors always use time-delay action fusing devices such as circuit breakers, time-delay Edison base fuses, or have overcurrent devices built into the motor, Figure 3. The properly rated GFCI can also be used for protecting motors.

OVERLOAD CURRENT PROTECTION FOR RESISTANCE LIGHTING AND HEATING

On electrical equipment and appliances that are resistance loads, such as lighting and/or heating, fusing can be done with either the time-delay or the single element type fuse illustrated in Figure 3.

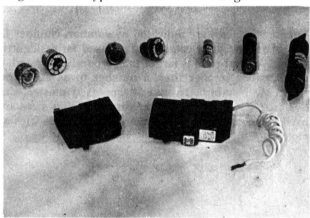

Fig. 3. Types of Fusing Devices.

CHARACTERISTICS OF OVERCURRENT DEVICES

As indicated earlier, overcurrent devices are designed to protect the circuit and its components including equipment operated in the circuit. The question could be asked in regard to how well an overcurrent device really protects a circuit. The one major factor is the time to trip or open the circuit.

Data in Table 1 reveal the time it takes to trip an overcurrent device based on the percent rated current. Further a comparison is shown between the typical time-delay and non-time delay devices. For example, if a circuit is designed and fused for a 15 ampere load and a load of 15 amperes is applied to the circuit, neither the time-delay or the non-time delay device will trip. A **15 ampere load** on a **15 ampere circuit** would load the circuit to **100 percent** of its rated load. Using Table 1 locate 100 percent on the base line and follow the line up the graph. Note that the line does not cross either of the fuse lines.

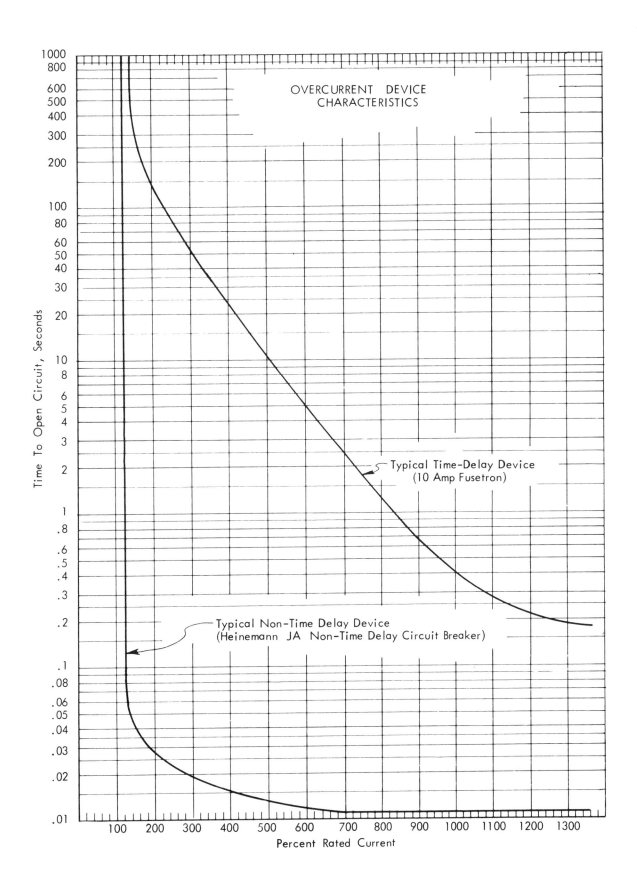

Table 1. Overcurrent Device Characteristics

Study Table 1 and solve this example. The previous **15 ampere circuit** has a **load of 45 amperes** applied to the circuit. If fused with a standard fuse, non-time delay, the fuse would trip in approximately **0.02 seconds.** If fused with a time-delay type fuse, the circuit would open in **50 seconds.** Forty-five amperes is **300 percent** rated current for a 15 ampere circuit; therefore, locate 300 on the base line and follow it up to the curved line. Trace across to the vertical line at the left and read the seconds to open the circuit.

A **short circuit** might be considered to load a circuit at **10 times** its rated ampacity. If a load of **150 amperes** were applied to a circuit having a rated ampacity of 15 amperes [**150 amperes divided by 15 amperes times 100 equals 1000 percent rated current**], the typical non-time delay fuse would blow in **0.011 seconds.** A time-delay device would trip or open the circuit in **0.4 seconds** under the same short circuit.

RESIDENTIAL AND FARM CIRCUITS
Circuit fusing is dependent on the purpose for which the circuit is designed. Circuits for **residential wiring** can be broken down into three types, namely:
- **General Purpose Circuits**--Circuits designed mainly for lighting and convenience outlets, 120 volts only
- **Appliance and Laundry Circuits**--Circuits used in the kitchen and laundry areas of the home, 120 volts only
- **Special Purpose Circuits**--Circuits to which only **one** appliance, such as a water heater; is connected, either 120, 120/240, or 240 volts

Circuits for **agricultural wiring** can also be broken down into three types, namely:
- **Lighting and Convenience Outlet Circuits**--120 volts only
- **Branch Circuits**--Circuits to which multiple electrical units are connected such as motors, and are too large for the lighting and convenience outlet type of circuit, 120 or 240 volts
- **Special Purpose Circuits**--Circuits to which only **one** apparatus is connected, either 120, 120/240, or 240 volts

The minimal diameter of copper wire used for electrical conductors vary among the residential and agricultural wiring circuits. **Wire size** is indicated by an **American Wire Gage [AWG] number.** General purpose residential circuits may be wired with number 14 or larger diameter wire conductors. The appliance circuit in the residence and the branch circuits of agricultural buildings must be done with number 12 or larger. Special purpose circuits must be number 12 or larger and the load in amperes will dictate the minimum size of wire conductors. In fusing these circuits there is a safe maximum that should be used for each wire size. The common wire sizes and maximum fuse sizes are:

AWG Wire Size	Maximum Fuse Size
14	15
12	20
10	30

Basically, this means that the appliance circuit in the home, for example, should be number 12 wire conductor, and no larger than a 20 ampere fusing device can be used to protect the circuit. However, there is nothing wrong with using a 15 ampere fuse on this circuit just to **play it safe** and even further protect refrigerators and other appliances commonly connected to the appliance circuit. The purpose of a fuse is to (1) protect and (2) to **blow** or otherwise disconnect the **hot** source conductor when something goes wrong such as a short circuit or a current overload. The fuse is therefore a safety valve and it warns us when something is wrong.

CLASSROOM EXERCISE V

PROVIDING ADEQUATE WIRING AND OVERLOAD PROTECTION

1. Most electrical conductor wires are made from what type of material?
   _____

   A. Gold   B. Silver   C. Copper   D. Aluminum

2. Place these AWG wire sizes from smallest to largest cross-sectional size: 6, 10, 4, 24, 18, 0.

   _____   _____   _____   _____   _____   _____
   Smallest   Largest

3. Define ampacity: _____
   _____

4. What is voltage drop? _____
   _____
   _____

5. Compute the percent voltage drop on this circuit. _____ percent.

   [Diagram: Source with Voltmeter = 236 on left, Voltmeter = 225 on right]

6. Allowable voltage drop for most home and farm wiring circuitry design is _____ to _____ percent.

7. Give several advantages to underground wiring versus overhead in air wiring.
   A. _____

   B. _____

8. Why is it so important that an unattended electric motor be properly fused?
   _____
   _____

9. Check which of the following overload current protectors have time-delay action.

　　____ Circuit Breaker　　　　　　　　　　　　　　　　　____ Cartridge Fuse

　　____ Standard Single Element Edison Base Fuse　　　____ Dual Element Edison Base Fuse

　　____ Motor Overload Protector

10. An appliance circuit must be wired with at least a number _____ AWG conductor size and protected with a fuse size no larger than _____ amperes.

11. Special purpose circuits are fused according to _____ with a minimum wire of AWG number _____ .

12. Using the Tables in the Appendix, determine the wire size to use for the following situation when limiting voltage drop to two percent:

Distribution Center — 75 feet — 3/4 hp motor (Use 125% Full Load Amperes)

Distribution Center — 90 feet — Small Building with 3000 Watt load

Type of Run	Motor (120V)	Motor (240V)	Building (120V)	Building (240V)
Underground or Conduit				
Overhead in air-one span				
Overhead in air-two spans				

LABORATORY EXERCISE VI

PROVIDING ADEQUATE WIRING AND OVERLOAD CURRENT PROTECTION

DEMONSTRATIONS

Purpose: To learn the importance of adequate wire size and proper protection with fuses.

Equipment:

1. Disconnect box: 30 amperes connected to 120/240V service
2. Receptacles: Five porcelain
3. Lamp: One 40W, 120V
4. Fuses: Three, 15 ampere, standard; one, 30 ampere, standard; and, one 10 ampere, time-delay.
5. Heater elements: Two, 600W, 120V; and two, 1000W, 120V
6. Ammeter: AC, 0-15 scale
7. Voltmeter: AC, 0-150 scale
8. Dropcord: 100 feet of AWG #18 with center tap for obtaining 50 feet
9. Conductor: 50 feet of AWG #14 copper wire.
10. Conductor: 2 feet of AWG #14 copper wire
11. Conductor: Short piece of approximately AWG #26 copper wire with no insulation
12. Motor: 1/3 hp split-phase motor with spring scales to measure locked rotor pull

Procedure:

1. **Overload Circuit**--Place two 1000W heater elements; two, 600W heater elements; and a 40W lamp in one side of the 120/240V circuit. Install a 15 ampere standard fuse in the circuit and operate.

   a. Results:

   _____

   _____

   _____

   b. Why did this happen?

   _____

   _____

2. Install a 30 ampere fuse in the circuit, but use a 6-inch piece of bare wire of approximately AWG number 26 between the disconnect box and the load on the 'hot' side of the circuit. Place a piece of tissue paper over the bare wire.

   a. Results:

   _____

   _____

   b. Explain why:

   _____

   _____

   _____

   _____

3. Replace the number 26 bare wire with a standard insulated wire. Remove one 1000W heater element and place it in the other side of the 120/240V circuit. Protect both sides of the 240V circuit with a 15 ampere standard fuse.

   a. Results: _____

   _____

   _____

   b. Why did neither fuse blow?

   _____

   _____

   c. Were the two 120V circuits within the 240V circuit balanced?
      How might the circuits be more equally balanced? _____

   _____

   _____

   _____

   Draw a diagram to illustrate this and label conductor colors.

4. **Voltage Drop Determination**--Provide experimental information to columns (a), (b), and (c) in the voltage drop chart by conducting four separate trials. All trials have the same load of one, 40W lamp; one, 600W heater; and one, 1000W heater in a 120V circuit with ammeter connected as shown. Use the flexible leads from the voltmeter for determining volts in different locations in the circuit, as directed by the trials.

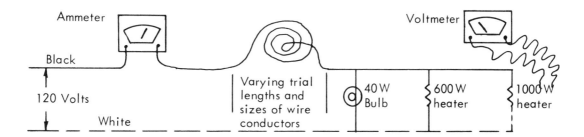

Trial 1 - Connect 2 feet of AWG number 14 wire in the 'hot' or black wire side of the circuit between the source and 40W lamp. Check volts at disconnect box and at the 1000W heater (load); check amperes in the circuit. Record results in chart.

Trial 2 - Replace the 2 feet of AWG number 14 wire with 50 feet of AWG number 14 wire and record results.

Trial 3 - Replace the 50 feet of AWG number 14 wire with 50 feet of AWG number 18 wire and record results.

Trial 4 - Replace the 50 feet of AWG number 18 with 100 feet of AWG number 18 wire and record results.

VOLTAGE DROP CHART

AWG No.	Conductor length	Volts at Box (a)	Volts at Load (b)	Amps (c)	Watts Box Volts x Amps (d)	Watts Load Volts x Amps (e)	Watts lost (d-e)	Cost of watts lost during a 24-hour period at 3¢ per kWh (f)
14	2 feet							
14	50 feet							
18	50 feet							
18	100 feet							

  a. Feel the 100 feet of AWG number 18 dropcord wire. Does it feel warm? _____
    Why? _____
    _____

  b. For each trial multiply factors to determine watts in columns (d) and (e). Then subtract (e) from (d) to obtain watts lost due to inadequate wiring. Calculate the cost of the loss as directed in column (f).

c. As wire length increased, (holding wire size constant), voltage drop _____.

d. As diameter of wire decreased, (holding wire length constant), voltage drop _____.

e. As wire length increased and diameter of wire decreased, amps flowing in the circuit generally _____. Why?

_____

_____

5. Place only a 40W lamp in a 120V circuit that has the 100 feet of AWG number 18 cord in it. Take voltage reading at lamp and disconnect box.
   a. Results:
      _____ Volts at disconnect box.   _____ Volts at lamp.

   b. Why was the voltage drop small? _____

6. Place a 1/3 hp split-phase motor (locked to a 2½-inch diameter pulley attached to a cord leading to a spring scale) in a 120V circuit containing 2 feet of AWG number 14 wire. Protect motor with a 10 amperes time-delay fuse. Operate the motor for only a few seconds, to avoid damaging motor, and record pull exerted on scales.   _____ Pull in Pounds

   Remove the 2 feet of wire and replace with 50 feet of AWG number 14 wire.
   Record Pull.   _____ Pull in Pounds

   Replace the 50 feet of AWG number 14 wire with 50 feet of AWG number 18 dropcord and record pull.   _____ Pull in Pounds

   Replace the 50 feet of AWG number 18 wire with 100 feet of AWG number 18 dropcord and record pull.   _____ Pull in Pounds

   a. Generally, electric motors have the best starting torque when wire length is _____ and wire diameter is _____.

   b. Why would a good wiring situation, compared to a poor one, improve motor life?

   _____

   _____

   _____

NOTES

TRANSMISSION AND DISTRIBUTION OF ELECTRICITY

FROM ENERGY TO ELECTRICITY

The electrical energy or power that is available to the consumer is transmitted and distributed through a network of **power lines** and **transformers**. Electricity starts at the generating station. The generating station is an energy conversion plant, meaning one form of energy (for example, coal) is burned and converted by mechanical means to electricity. Following is a list of power sources used to make electricity:

Table 1. Power Conversion to Electricity.

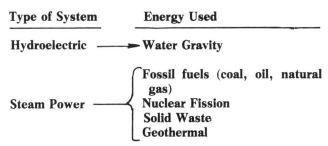

Water gravity or hydroelectric plants use the force of falling water from one elevation to another to turn electrical generators. All other types of generating plants heat water to steam and the steam produced by heating water turns steam turbines connected to generators to produce electrical power. Three of the four types of steam power systems use unrenewable natural resources. Only geothermal uses the steam of natural geysers to produce electricity from a basic renewable source. However, only a few sites exist on the face of the Earth where this can be done on a practical basis. The unrenewable fuels of coal, oil, natural gas, nuclear fuels and solid waste produce most of the electricity in North America. The hydroelectric generating plants produce electricity with no fuel, but similar to the geothermal power source, there is a limit to its application. Only a relatively few rivers in the U.S. can be dammed on a practical basis to produce hydroelectric electricity.

GENERATORS

Generators are connected to turbines or other mechanical power units that turn them. The rotating action of the generators produce the electricity that is ultimately used in homes, farms and industry. Generators produce electricity by means of a principle discovered in 1832 by Michael Faraday, an English physicist. Faraday found that he could produce electricity in a copper wire by moving the wire near a magnet, or by moving a magnet near a wire. This process is called **electromagnetic induction.** The voltage, or electromotive force (**Emf**) of the electricity produced in this way is called **induced voltage** or **induced electromotive force.** If the copper wire is part of a closed circuit of wires, the induced voltage causes electric current to flow to the circuit. This principle is the basis of all generators used today, whether small portable generators powered by gasoline engines or giant generators powered by steam turbines.

Two types of currents can be produced from generators; **A-C(alternating current)** or **D-C** (**direct current**). The electricity used in our homes is the A-C type and has a **Sine wave** as pictured in Figure 1.

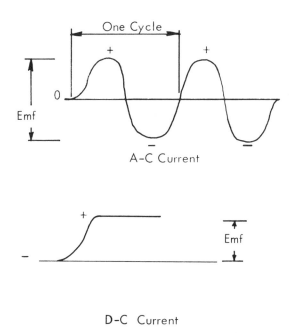

Fig. 1. Sine Waves of A-C and D-C Power Sources

The A-C current is alternating 60 times per second. This means the typical lamp is going **off** and **on 120 times per second** and is called **60 cycle current;** however, it is not noticeable to the human eye. The flashlight and automobile electrical systems use D-C current with the power source being a drycell or wet storage battery. Its Sine wave is also illustrated in Figure 1. When the flashlight is turned on, the voltage or Emf potential moves to the extent of its potential electromotive force and is relatively stablized to discharge its force through the lamp resistor. Therefore, D-C current does not alternate. The

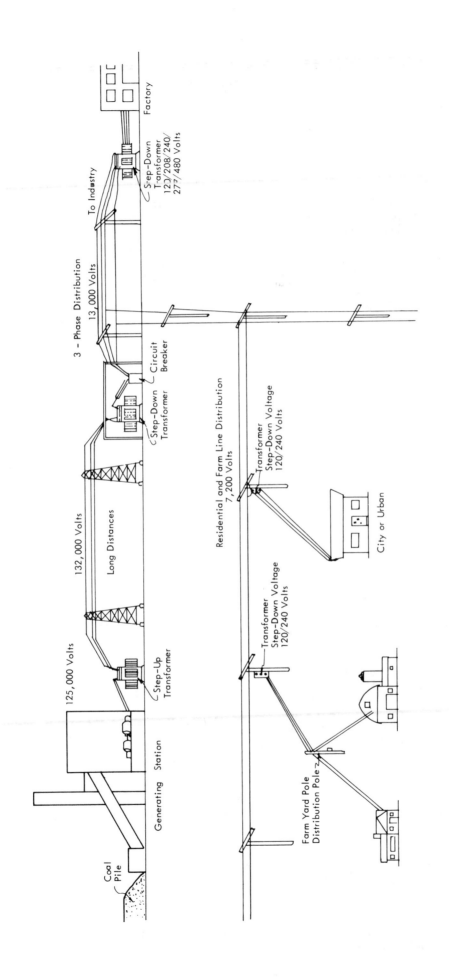

Fig. 2. Production, Transformation and Distribution of Electricity.

alternation of the A-C current is used to an advantage in the operation of many electric motors, especially motors where speed is governed by the cycles per second alternations.

For all practical purposes the electrical power and distribution system, from generating plant to the kitchen light, is A-C current. However, the distribution system is not quite that simple and requires an explanation.

THE TRANSFER OF ELECTRICITY

Generating electricity is only part of the process of supplying electric power. The electricity must be transmitted from power plants to the city or areas to be used. Then the electricity must be distributed to homes, farms and factories. The network of electrical power transferred from the generating station to consumer is shown in Figure 2. The voltage of the transmission lines may vary considerably from one power supplier to another, especially at the generating station source. Power suppliers find it advantageous to transmit electrical power over long distances at very high voltages because voltage drop (a waste) is greatly reduced when amperage (flow of current) is lowered.

TRANSFORMERS

Transformers are used to either **step-up** or **step-down** electrical power. When a group of transformers are placed at one site, the transformer complex is called a sub-station. Step-up transformers receive low voltage and high amperage and change it to high voltage and low amperage. Step-down transformers do the opposite as shown in Figure 3. The illustrations assume the transformers are 100 percent efficient. Of course, they are not as there is some loss of power in the transforming process. The number of **coil windings or loops** in the **primary** and **secondary** sides determine the step-up or step-down characteristics of the transformer. For example, the step-down transformers near the residence has more windings on the primary side and fewer windings on the secondary side. This transforms **high voltage (7200 volts)** to the **lower voltages (120/240 volts)** used in homes.

The formula for determining the primary side to secondary side relationship of a transformer is as follows:

Voltage on Primary Side Times Amperage on Primary Side = **Voltage on Secondary Side Times Amperage on Secondary Side**

For example, if the secondary side voltage is 240, primary side voltage is 7200, and 75 amperes are flowing in the secondary side, then 2.5 amperes will be flowing in the primary side of the transformer, because:

Primary		Secondary
7200 V x A	=	240V x 75A

7200A = 18,000

$A = \dfrac{18,000}{7200}$

A = 2.5 Amperes on Primary Side

Transformers on power distribution systems are the property of power suppliers. For safety sake, the layman should **never** attempt to work on or repair transformers as they carry high voltage and high amperage making them very dangerous.

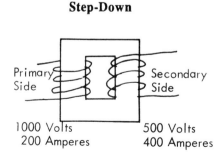

Fig. 3. Step-up and Step-down Transformers.

CLASSROOM EXERCISE VI

TRANSMISSION AND DISTRIBUTION OF ELECTRICITY

1. Electrical power is transmitted and distributed through a network of _____ and _____ .

2. When water is used as an energy source to turn generators, the type of electric producing system is called _____ .

3. Name two of the three commonly used fossil fuels for producing electricity? _____ and _____ .

4. Generators produce electricity by means of a principle discovered by _____ .

5. The two types of currents produced from generators are _____ and _____ .

6. _____ is greatly reduced when electrical power is transmitted over long distances at high voltages.

7. What determines whether a transformer is a step-up or step-down? _____
   _____
   _____
   _____

8. Compute the expected secondary side amperage (assume 100% efficiency) for a step-down transformer that has 600 volts and 200 amperes on the primary side when the secondary side voltage is 400. _____ Amperage.

NOTES

THE SERVICE ENTRANCE

The **service entrance** is the part of the electrical system that transfers electricity from the power supplier's transformer to the service entrance panel. Illustrated in Figure 1 is a typical single phase entrance feed from the transformer through a service drop (the main set of cables to the house of building), into the weatherhead, down through a conduit (or cable), through the kilowatt-hour meter, and into the **main disconnect** of the entrance panel. The ground rods at both the transformer pole and entrance panel are also an important part of the system.

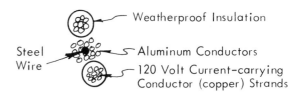

Fig. 2. Cross-sectional View of a Triplex Cable.

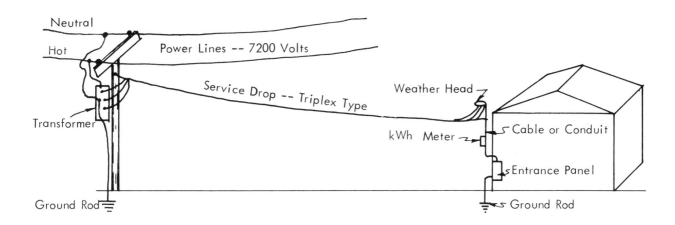

Fig. 1. Transformer to Entrance Panel Service Drop Feed.

OVERHEAD IN AIR

The **service drop cable** shown in Figure 1 is a **triplex cable**, most commonly used today for residential applications. It consists of two insulated current-carrying conductors spiralling around a non-insulated cable. The non-insulated support cable is made of steel and aluminum--steel for strength and aluminum for conductance. The support cable is also the **neutral** ground going from the transformer to the entrance panel inside the building. A cross-sectional view of the triplex cable appears in Figure 2. The triplex support cable physically holds up the weight of itself and the two spiralling conductors by means of fasteners securely affixed to the power pole and to the house or building.

The big advantage of the triplex type service drop is that it is quite strong, compared to running three separate conductors. This strength is important regarding wind storms, ice storms or other natural hazards. Going underground with the service feed is the best protection against the natural elements and hazards. The underground system will be discussed later in this unit.

Older-type systems use the same three conductor cable drop. However, the three conductors are run separately and each securely affixed with an insulator at both ends of the run, as shown in Figure 3. One disadvantage of this older system is that all three conductors must physically hold themselves and withstand the elements of nature, especially wind and ice problems. The size of conductors has to be of appropriate size to make the run and handle the load. From a practical standpoint, conductors in overhead spans must be at least AWG number 10 for spans up to 50 feet and number 8 or larger for longer spans.

UNDERGROUND

The **underground burial system**, such as illustrated in Figure 4, is rapidly becoming popular. Used in the underground manner, the feed is called a **service lateral** rather than a service drop. It has these **advantages** compared to overhead in air service drop systems.

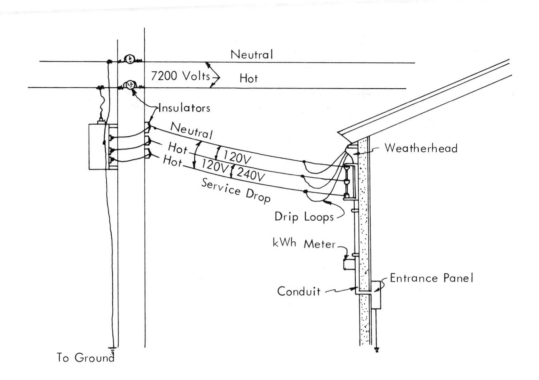

Fig. 3. Single Conductors Used as Service Drop Feeders in the Single-phase 120/240 Volt System.

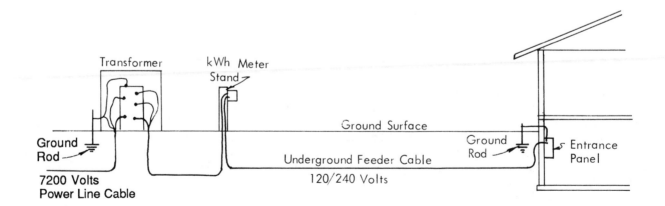

Fig. 4. Typical Residential Underground Installation.

- **Safety:** Accidental contact to current-carrying conductors is greatly minimized, as with metal ladders, agricultural equipment, children flying kites, and similar accidental hazards due to conductors overhead.
- **Weather:** The elements virtually can not damage the system.
- **Long Life:** The installation should last many years longer and essentially be maintenance free.
- **Location:** It allows the power supplier to locate transformers and metering equipment close to electrical load centers without exposing premises to high voltage line hazards.
- **Esthetic Value:** The esthetic value or beauty of the installation is improved. Any equipment above the ground (transformers and meter stands) can be hidden with planting of shrubbery or with attractive security fences.

Disadvantages of the underground feeder service laterals are:

- **Cost:** Cost of installation, including trenching and special underground feeder lines which cost more than conductors used with overhead service.
- **Flexibility:** If the system needs to be changed to larger conductors because of added loads, the wires must be taken up and replaced or abandoned, unless the underground feeder conductors are in a raceway, duct or conduit.
- **Safety:** There is always a possibility of digging into an underground system, unless workers have precise knowledge of where the cables are buried.

TYPES OF CABLES AND CONDUCTORS FOR SERVICE ENTRANCES

A service entrance is the non-fused portion of the run from transformer to main entrance panel disconnect. **Type SE (Service Entrance)** can **not** be used underground but is most commonly used in home and agricultural wiring from the weatherhead mast down the side of the building through the building wall and to the **main** of the entrance panel disconnect. Type SE is usually an insulated cable composed of stranded and shielded wires surrounding two insulated current-carrying conductors. The current-carrying conductors are also stranded as shown in Figure 5.

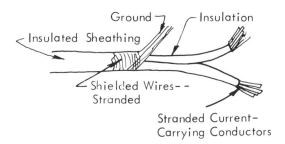

Fig. 5. Type SE Multiple Conductor Service Entrance Cable Used for 120/240 Volt Service.

Type USE (Underground Service Entrance) is commonly used to make the run underground from the transformer to residence or other buildings as shown in Figure 4. It may be direct burial or run through metallic or nonmetallic ducts or raceways such as galvanized steel conduit, polyvinyl chloride tubing, high density polyethylene tubing, asbestos pipes, concrete raceways and other rigid approved materials. Local code provisions govern rules and regualtions on such installations.

Type UF (Underground Feeder) may **not** be used as a service entrance, but only as a branch circuit conductor or feeder. Feeders are conductors leading from one over-current protector (fusing device) to another over-current protector.

All three types of previously mentioned cables (SE, USE and UF) may be a single insulated conductor or multiple conductors. When individual conductors are used overhead in air, as in Figure 3, the **Type WP (Weather Proof)** is most commonly used; however, Type USE may also be used, but not Type UF. Type UF should **not** be exposed to sunlight.

When types USE or UF are placed underground and not in a conduit or raceway, they should be buried at least **24 inches** below ground surface or deeper where conditions such as deep cultivation or trenching might take place. It is recommended to lay conductors in the trench cushioned by several inches of sand base both below and above the conductors before back filling with soil. Also, the conductors should not be pulled too tightly, but laid in a non-tensioned manner to allow for expansion due to natural soil disturbances. Most importantly on underground installations, a detailed map showing precise placement of all cables and conductors should be made, filed for safe keeping and referred to for future excavations in the area.

THE ENTRANCE PANEL

Service entrance conductors terminate in the **entrance panel**. All circuits in the building start at the entrance panel and branch out through various circuits, such as those mentioned in Unit VI. Many common terms are used to describe the entrance panel; among those are **circuit box, circuit breaker box, distribution panel, entrance panel, service box, fuse box and branch circuit box.**

The modern entrance panel is usually wired to the three-wire system as discussed in the first part of this Unit. Therefore, 120/240 volts are connected to the main of the entrance panel as in Figure 6, illustrating a small entrance panel of low amperage rating.

The entrance panel has at least **three** main parts:
- The main switch or disconnect
- Fusing or overcurrent protection devices
- Terminals for attaching branch circuit distribution conductors

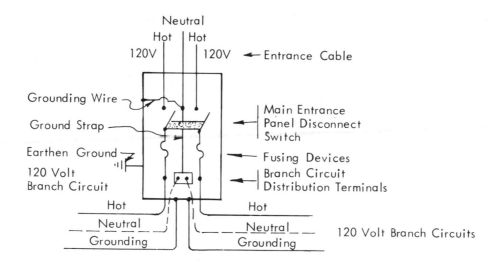

Fig. 6. A Small Entrance Panel Connected for 120/240 Volt Service.

The entrance panel illustrated in Figure 6 has all three of these parts. Note that the switch, (**main disconnect**) located between the entrance cable and fuses, is a double-pole, single-throw switch and is shown as open. When the switch is closed, current can flow through the fuses and to the two branch circuits. The neutral of each branch circuit is connected to a **neutral strip** leading directly, unfused and unswitched to the entrance cable neutral. Furthermore, this neutral strip is grounded to the metallic entrance panel box with a grounding wire, strap or bonding screw, and the box is grounded to earth. Therefore, the entrance cable neutral, the neutral strip, and the metallic box are all in **electrical continuity** or **bonded** together to earth through a driven ground stake as near as practical to the entrance panel. A single, solid bare copper conductor is used to go between the entrance panel and the driven stake.

Entrance panels used in the home are generally larger than the one illustrated in Figure 6. Size of entrance panels are rated by the **ampere load** and common sizes are **30, 60, 100, 150 and 200**. Homes must have at least **100 ampere entrance panel** according to NEC requirements. Larger entrance panels are available and used in industrial applications, and some agricultural situations where loads are above 200 amperes. In these latter cases, three-phase current is commonly used and there are three main fusing devices protecting each **hot** service line conductor, and the main disconnect switch is a triple-pole, single-throw switch. A 100 ampere entrance panel with six branch circuit plug-type fuses is shown in Figure 7.

The fusing and wiring schematic for such a 60 ampere entrance panel, including the range is illustrated in Figure 8. In this type of entrance panel, all current must go through the **main fuses** or **overcurrent**

Fig. 7. A 100 Ampere Entrance Panel with Plug-type Branch Circuit Fuses.

protectors. The main disconnect fuses or service breaker are therefore rated larger than any other fuses found in the system and could have up to 60 ampere fuse or breaker disconnect since it is a 60 ampere rated entrance panel. Common **service disconnect sizes** are **30, 40, 50, 60, 70, 100, 125, 150, 175 and 200 amperes.** The fuses used in the range side of the panel would probably be 30, 40 or 50 depending on the ampere rating of the range or any other 240 volt special purpose appliance. Note the sub-feed at the bottom of the schematic. Some entrance panels have this sub-feed to connect another, but smaller, rated electrical panel. For example, the connection of a 30 ampere rated panel protecting a 240 volt water heater might be made from this sub-feed. No intent was made to show the **neutral** ground and the grounding system in Figure 8, but it would be similar to the situation explained in Figure 6.

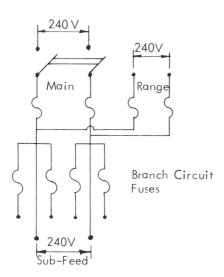

Fig. 8. Wiring and Fusing Schematic for a 60 Ampere Panel with Sub-feed. No Neutral Grounding System Shown.

Entrance panels use overcurrent protection devices as presented in Unit VI. When cartridge-type fuses are used as the main, the removable fuses also act as the switch. The cartridge fuses are mounted in an **insulated block** which is a **pullout** as shown in Figure 9. The main switch is opened by pulling the block out, and all current to all parts of the entrance panel is stopped.

Fig. 9. A 100 Ampere Entrance Panel with Pullout Main and 6 Plug-type Branch Circuit Fuses.

Circuit breakers, used as **mains** and also to service individual branch circuits, are becoming more popular for use in entrance panels. An entrance panel typical of a modern home is shown in Figure 10.

Fig. 10. Typical Entrance Panel with Circuit Breakers.

Circuit breakers allow the user to reset the breakers when tripped by an overload or **shorted** circuit. Whereas, plug-type or cartridge-type fuses must be replaced when they are **blown**. Since the ground-fault circuit interupter is required in most residential installations, the type manufactured for use in entrance panels is shaped and used similarly to a circuit breaker.

SELECTING THE SERVICE ENTRANCE PANEL AND DISCONNECT

Size of the entrance panel is determined by the anticipated load or demand. A home with electric heat will require a larger entrance panel than a home not having electric heat. Some general rules of thumb are presented for sizing and selecting an entrance panel.

Following are rules and recommendations for calculating the service entrance panel of a home **without electric heating**:
(1) **General purpose lighting and outlets:** Use 3 watts per square foot of living space.
(2) **Appliance or kitchen circuits:** There shall be at least 2 appliance circuits with 1500 watts per circuit used.
(3) **Laundry circuit:** Figure 1500 watts for the laundry circuit.
(4) **Range:** Use 8000 watts for ranges up to 12,000 watts. For Larger ranges use 400 watts per 1000 watts above 12,000, and add to the original 8000.

(5) **Special purpose permanently fixed appliances:** If 3 or fewer are listed use 100% of the wattage value of the 3 appliances. If 4 or more are listed use 75% of the wattage value of all appliances listed. See Table 8 in the Appendix for wattage rating of various appliances.

Compute the total wattage by adding the wattages of (1) plus (2) plus (3). Then take 100% of the first 3000 watts and 35% of the remainder. Add this to the wattage of (4) and (5). For purposes of explanation, the following example is given for a home **without electric heat**, following the recommendations previously stated:

(1) **General purpose lighting and outlets:**
   1200 sq. ft. x 3 watts/sq. ft. = 3,600
(2) **Appliance or kitchen circuits:**
   2 at 1500 watts/circuit = 3,000
(3) **Laundry circuit:**
   1 at 1500 watts/circuit = 1,500
   = 8,100

   First 3000 watts at 100% = 3,000
   8,100-3,000 = 5,100 x 35% of remainder = 1,785
(4) **Range:** Assuming at 14,000 watt range = 8,800
(5) **Special purpose permanently fixed appliances:**
   a. Space heating or air conditioning.
      Assume space heating (gas fired furnace is 600 watts and air conditioning (window unit) is 2,000 watts. The larger should be used. = 2,000
   b. Water heater = 4,500
   c. Clothes dryer = 4,000
   d. Dishwasher = 600
      11,100

   Since there are 4 special purpose permanently fixed appliances, 75% will be used. = 8,325

   Total Computed Load------21,910

Divide by 230 volts to obtain service entrance panel size.

$$\frac{21{,}910 \text{ Total Computed Load}}{230 \text{ Volts}} = 95 \text{ amperes}$$

Therefore, a **100 ampere entrance panel** would be needed with a **service disconnect** also of **100 amperes rating**. If future expansion is anticipated, a 200 amperes entrance panel could be selected, but a 100 amperes service disconnect would be used until additional load requirements dictated a larger one of 125, 150, 175 or 200. These larger ones could be placed into a 200 ampere entrance box.

Following are rules and recommendations for calculating the service entrance panel of a home **with electric heating:**
(1) **General purpose lighting and outlets:** Use 3 watts per square foot of living area.
(2) **Appliance or kitchen circuits:** There shall be at least 2 appliance circuits with 1500 watts per circuit used.
(3) **Laundry circuit:** Figure 1500 watts for the laundry circuit.
(4) **Range:** Use actual wattage rating of the range.
(5) **Special purpose permanently fixed appliances:** List all at actual wattage.
(6) **Electric home heating:** If 3 or fewer electric heating units, separately or commonly controlled, use 100% demand wattage. If 4 or more units, separately controlled, use 40% demand wattage. General rules of thumb for house heating is 7 to 12 watts per square foot of living area, depending on climatic conditions. Ten is a good estimate for most parts of the U.S.
(7) **Air conditioning:** Usually air conditioning units, either central or window units, are using fewer watts than the electrical heating units. Since they would never be 'on' at the same time, the electric heat wattage is recorded in (6) above and air conditioning is zero watts when heating wattage is higher than air conditioning. If air conditioning is higher, then it is recorded and heating is zero.

Add (1) through (5) and use 100% of the first 10,000 watts. Then take 40% of the wattage for the remainder. Using directions of (6) and (7) above, add either 100% or 40% to that value for obtaining total computed load. Several examples for purpose of explanation follows regarding calculating the service entrance panel for homes with electric heating. The loads will be the same with the exception of the electric heating units:
(1) **General purpose lighting and outlets:**
   1200 sq. ft. x 3 watts per sq. ft. = 3,600
(2) **Appliance or kitchen circuits:**
   2 at 1500 watts/circuit = 3,000
(3) **Laundry circuit:**
   1 at 1500 watts/circuit = 1,500
(4) **Range:** Assuming a 14,000 watt range = 14,000
(5) **Special purpose permanently fixed appliances:**
   a. Water heater = 4,500
   b. Clothes dryer = 4,000
   c. Dishwasher = 600

Total Load Excluding Electric Heat --- 31,200
   First 10,000 watts at 100% = 10,000
   31,200 - 10,000 = 21,200 x 40% = 8,480
(6) **Electric home heating:** Assume 10 watts per sq. ft. or 12,000 watts.
   If **3 or fewer** controls, 100% **(12,000)**
   If **4 or more** controls, 12000 x 40% (4,800)

Total Computed Loads ---- [30,480] or [23,280]

For a home with 3 or fewer controls for electric heat would require an entrance panel of:

$$\frac{30{,}480 \text{ Total Computed Load}}{230 \text{ Volts}} = 133 \text{ amperes}$$

and a 150 ampere entrance panel would be selected along with a service disconnect of 150 amperes rating. The home with 4 or more electric heating controls would need an entrance panel of:

$$\frac{23{,}280 \text{ Total Computed Load}}{230 \text{ Volts}} = 101 \text{ amperes}$$

with **150 amperes rated entrance panel** selected along with a **service disconnect of 125 amperes rating**. Generally, most medium and large size homes with electric heat need a 200 amperes rated service entrance panel or box.

NOTES

CLASSROOM EXERCISE VII

THE SERVICE ENTRANCE

1. The service entrance transfers electricity from _____ to _____

2. Name four parts of the service entrance system.
   A. _____
   B. _____
   C. _____
   D. _____

3. A triplex service drop is very strong because of a _____ wire in the middle of its neutral conductor.

4. What is meant by the 120/240 volt system?

   _____
   _____
   _____
   _____

5. For a single conductor overhead in air, when spans are less than 50 feet, and AWG number _____ must be used, or an AWG number _____ for longer spans, even when loads are small.

6. List four advantages of underground installations compared to service drop installations above the ground.
   A. _____
   B. _____
   C. _____
   D. _____

7. List two disadvantages to underground feeder service laterals.
   A. _____
   B. _____

8. Type SE in regard to a cable stands for _____

9. Type UF stands for _____

10. Name the three major parts of the entrance panel.
    A. _____

    B. _____

    C. _____

11. Which of the following amperage ratings are common entrance panel sizes. Circle those that are correct.

    20   50   60   75   100   150   200   220   280

12. The common service disconnect sizes are _____
    _____ amperes.

13. Why is a stake grounding the entrance panel to the earth important?

    _____

    _____

14. All other fusing devices should be (smaller or larger) than the 'main' or disconnect fuses and the entrance panel rating.

15. Which type of entrance panels are becoming more popular?
    A. Those with standard plug type fuses
    B. Those with circuit breakers
    C. Those with cartridge type fuses

16. Compute the size of entrance panel needed for an 1800 square foot home without electric heat. Assume the home has a range of 11,000 watts, dishwasher of 800 watts, water heater rated at 4700 watts, 25 ampere -240 volt air conditioner, two appliance circuit and a laundry circuit. _____ Panel _____ Disconnect Needed

17. Assume the same home as above, but with electrical heating from six heating units each separately controlled. Assume 10 watts per square foot of living area.

    _____ Panel Needed; _____ Disconnect Needed

NOTES

PRACTICAL WIRING LABORATORY EXERCISES

INTRODUCTION AND SAFETY

The purpose of this unit is to learn to do practical wiring. Six laboratory exercises are presented that will acquaint the student with wiring situations commonly performed by electricians and other persons who may do some of their own wiring.

In wiring these exercises, one should always be concerned with **safety.** This includes the proper and safe use of tools while doing the exercises. The finished job should be a **safe electrical wiring job.** Therefore, when a wiring exercise is completed, ready for evaluation by the instructor, it should be:

(1) **Checked thoroughly with a continuity tester before plugging in the service power cord, or**
(2) **Plugged into a circuit protected by a ground-fault circuit interrupter [GFCI].**

One should never take chances around electricity. Remember that 120 volts is very dangerous, and can cause shock resulting in death or serious injury. If the wiring done on any of these exercises is incorrect, and wiring changes needed to be made, the **power cord** should **always be disconnected** from power source.

The service entrance in each of the six exercises has a type SJ, 14-3 power cord with 3-wire, 15 amp, 120 volt plug cap attached. The stranded wires of each conductor or the power cord should be **tinned** with solder to minimize loose wires. This also makes for a more secure connection or attachment to both the plug cap and the service entrance panel. Once the power cord is constructed, it can be used for other exercises.

TOOLS, PRACTICE WIRING PANEL AND MATERIALS

The following tools should be available for performing wiring exercises:

Wire Stripper
Cable ripper
Linemen'a pliers
Diagonal cutter
Long chain nose pliers
Screwdrivers
Electrical neon tester

A handy **electrical tool kit** with carrying box is available from Hobar Publications. It includes the tools listed above, and is shown in Figure 1.

The materials and **wiring practices panel** needed for completing the **six exercises** are shown in Figure 2. The panel and materials can be purchased from Hobar Publications and includes the necessary materials needed for doing the six wiring exercises on a constructed wiring panel. Also available from Hobar Publications are **two slide/cassette tape programs** that cover the details of completing these wiring exercises. One program is **Wiring a Switch & Lamp** and the other is **Wiring Three & Four-Way Switches.**

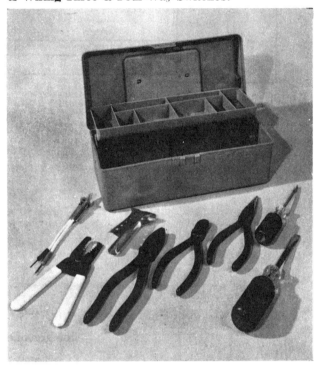

Fig. 1. Electric Tool Kit.

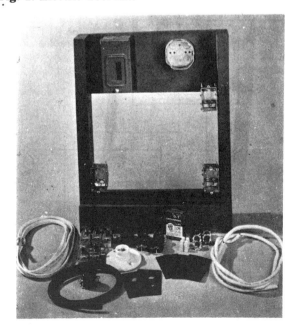

Fig. 2. Wiring Practices Panel.

LABORATORY EXERCISES VII

Wiring Practices Lab #1

Switched lamp with feed wire into lamp receptacle.

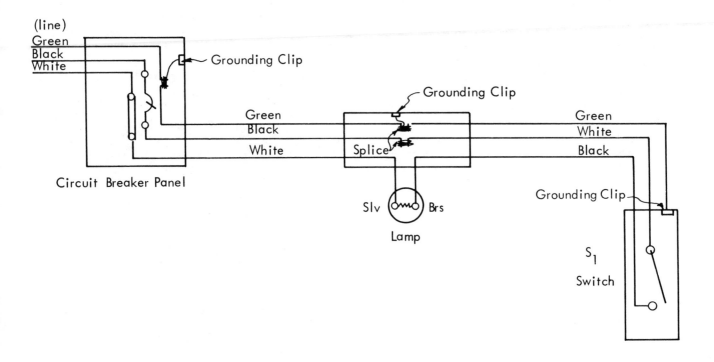

Wiring Practices Lab #2

Two grounded duplex receptacles.

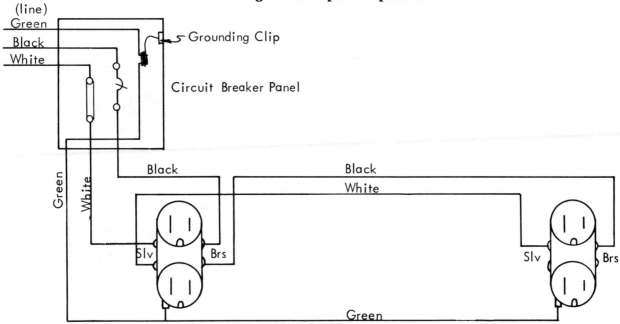

Wiring Practices Lab #3

Lamp switched by 2-S3 switches and 1-S4 switch with feed wire into lamp receptacle.

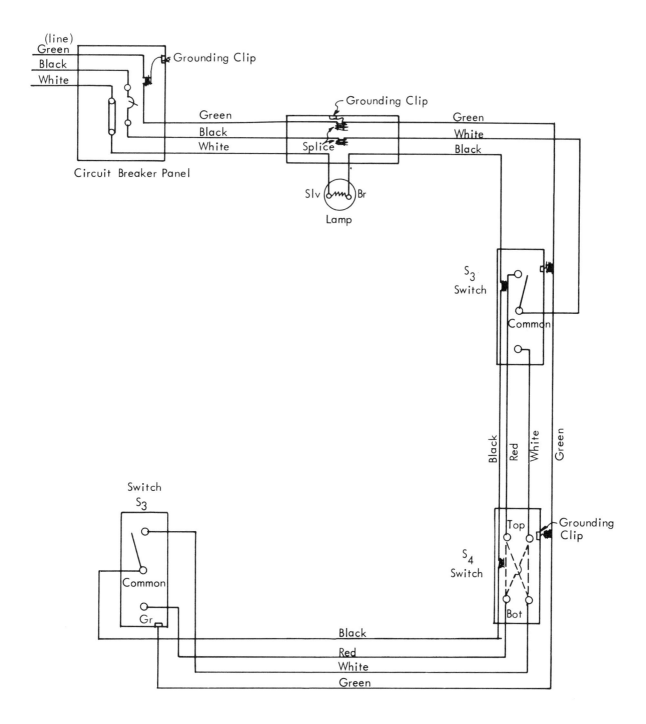

Wiring Practices Lab #4

Two grounded duplex receptacles, one switched with feed wire into unswitched receptacle.

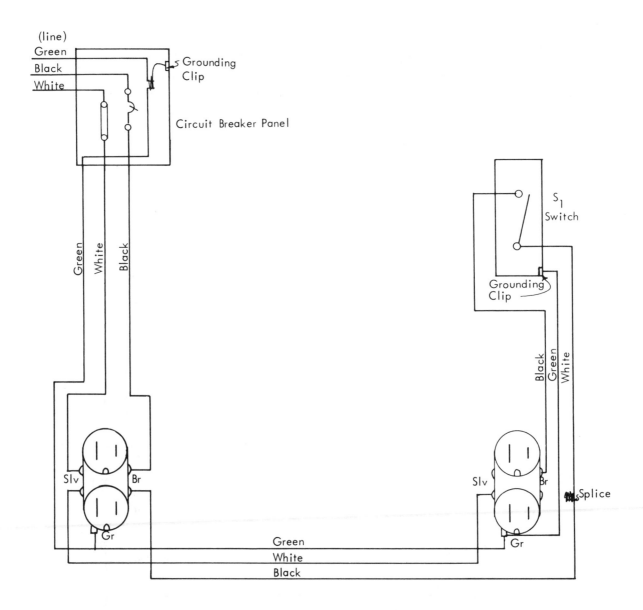

Wiring Practices Lab #5

Two grounded duplex receptacles and switched lamp with feed wire into lamp.

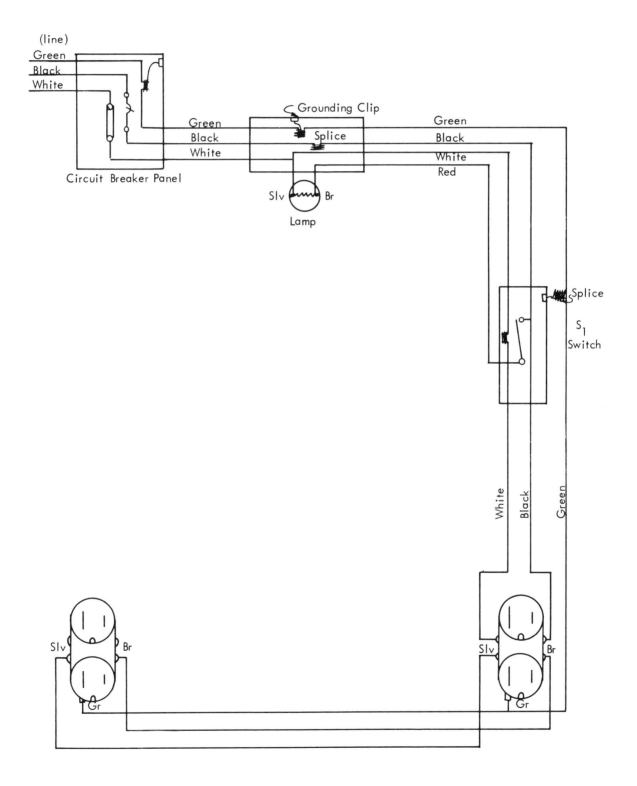

Wiring Practices Lab #6

Switched lamp and split receptacle, one side switched, with feed wire into lamp.

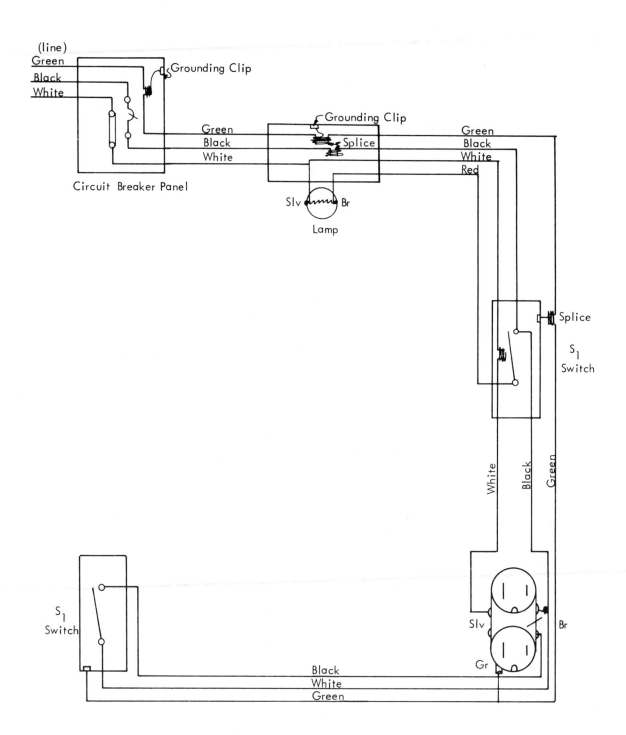

EVALUATION OF WIRING PRACTICES EXERCISES

The following evaluation form may be used to evaluate the completeness and quality of each of the six wiring exercises.

Item	Points Possible	Points Earned
Power cord	10	______
Service entrance correctly wired	10	______
Use of black and red **hot** conductors	15	______
Use of white **neutral** conductors	10	______
Use of 'grounding' conductors	10	______
Cable ripping	5	______
Wire stripping	5	______
Use of cover plates	5	______
Use of box clamps	5	______
Use of solderless connectors	5	______
Connecting conductors to terminals	10	______
Safety, use of tools, and work habits	10	______
	100	

Name (s) ____________________ Points ________

Exercise # ____________________ Date ________

Item	Points Possible	Points Earned
Power cord	10	______
Service entrance correctly wired	10	______
Use of black and red **hot** conductors	15	______
Use of white **neutral** conductors	10	______
Use of 'grounding' conductors	10	______
Cable ripping	5	______
Wire stripping	5	______
Use of cover plates	5	______
Use of box clamps	5	______
Use of solderless connectors	5	______
Connecting conductors to terminals	10	______
Safety, use of tools, and work habits	10	______
	100	

Name (s) ____________________ Points ________

Exercise # ____________________ Date ________

ELECTRICAL WIRING TOOLS

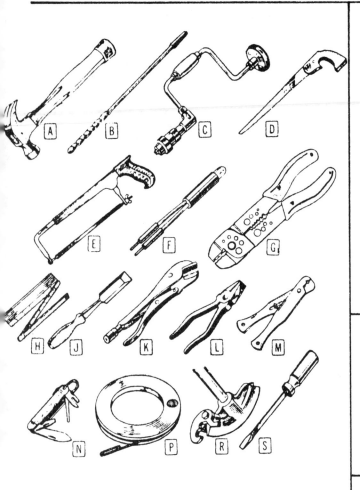

Operational Procedure:

1. Identify electrical tools by letter and give one specific use of each tool.
2. Tools _____, _____, and _____ would be used to cut insulation from conductors.
3. Tool _____ would be necessary if attempting to pull new cable between studs in an old building.
4. Before connecting an appliance to a circuit tool, _____ could be used to test the circuit.
5. Tools _____ and _____ would be used to put holes in floor joists to run wire under a floor.
6. Tool _____ would be best for cutting cable to length.
7. Tools _____ and _____ are necessary when doing conduit wiring.
8. Tool _____ could be used to crimp a spade lug to the end of conductor.
9. Perform electrical wiring jobs using tools.

Operation Teaches: Ability to ...

1. Identify electrical tools.
2. Determine common uses of electrical tools.
3. Understand importance of having proper tools for each job.
4. Use electrical tools.

Materials:

1. Electrical tools as illustrated.
2. Materials for using tools: 2X4's, cable, conduit, boxes, box connectors, and spade lugs.
3. Text: Basic Electricity and Practical Wiring -- Hobar #2377.

Identification:

	Name	Use
A.	_____	_____
B.	_____	_____
C.	_____	_____
D.	_____	_____
E.	_____	_____
F.	_____	_____
G.	_____	_____
H.	_____	_____
I.	_____	_____
J.	_____	_____
K.	_____	_____
L.	_____	_____
M.	_____	_____
N.	_____	_____
P.	_____	_____
R.	_____	_____
S.	_____	_____

Evaluation Score Sheet: Points

Item	Possible	Earned
1. Identification of tools	32	_____
2. Use of tools	32	_____
3. Questions 2-8	12	_____
4. Electrical jobs performed	14	_____
5. Safety and work habits	10	_____
Total	100	_____

Name _____

Date _____   Grade _____

MAKING A SOLDERLESS CONNECTION

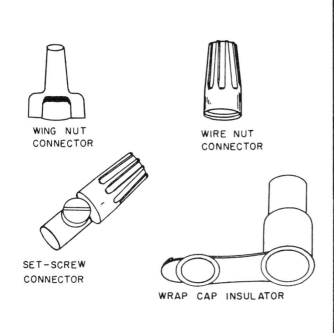

WING NUT CONNECTOR
WIRE NUT CONNECTOR
SET-SCREW CONNECTOR
WRAP CAP INSULATOR

Operation Teaches: Ability to ...

1. Understand different applications and where each type is used.
2. Identify types of connectors.
3. Determine size needed.
4. Apply connectors properly.
5. Carry out testing procedure.

Materials:

1. Lampcord AWG number 18, 1'
2. Nonmetallic cable 14-2, 2', (2-1" pieces)
3. Cable ripper
4. Wire stripper (1 per 2 students)
5. Miscellaneous connectors; various sizes

Name _____

Date _____ Grade _____

Operational Procedure:

1. Strip wire 1/8" less than length of connectors.
2. Connect 2 ends of No. 18 wire with correct size connector. Size Selected _____
3. Connect one end of #18 wire and black #14 with correct size connector.
4. Connect two white wires #14 with correct size connector. Size Selected _____
5. Remove connectors and check wire twist.
6. Replace connectors for grading.

Questions:

1. List the four main types of solderless connectors:
   1. _____ 2. _____ 3. _____ 4. _____
2. Why is wire size so important in connector selection? _____
3. Where would connectors normally be used in electrical wiring? _____
4. What are some alternatives to solderless connectors? _____
5. How much insulation should be removed for proper connector use? _____

Evaluation Score Sheet: Points

Item	Possible	Earned
1. Use of wire stripper	10	_____
2. Connection strength	20	_____
3. Wire equal twist	20	_____
4. Correct size of connector	10	_____
5. Insulation stripped correct distance	10	_____
6. Questions	20	_____
7. Safety and work habits	10	_____
Total	100	_____

MAKING A SOLDER CONNECTION

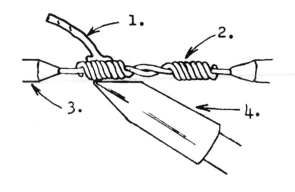

Part Identification:

1. _____
2. _____
3. _____
4. _____

Materials:

1. Electrical wire to be spliced
2. Rosin core solder
3. Soldering iron
4. Plastic electrical tape
5. Electricians pliers
6. Wire stripper

Operation Teaches: Ability to ...

1. Use electrical tools correctly.
2. Practice safe working habits with electrical wiring.
3. Splice a wire correctly.
4. Solder the splice correctly.
5. Tape the connection correctly.
6. Apply these skills to electrical wiring tasks.

Operational Procedure:

1. Remove insulation with wire stripper:
    a. Remove about 1 1/2 inches of insulation.
    b. Do not cut it off sharply.
    c. Avoid nicking the wire.

wrong   right

2. Twist wires together as shown:
    a. First step

    b. Finished joint

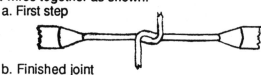

3. Soldering:
    a. Preheat soldering iron.
    b. Place soldering iron under the splice.
    c. Heat wire until it will nelt the solder.
    (Do not melt solder with the soldering iron.)

4. Taping:
    a. Insulation must be added equal to the original insulating value.
    b. Start taping at one end and wrap spirally toward the other end, then back and forth until the desired insulation is achieved.

Evaluation Score Sheet:

Item	Possible	Earned
1. Part identificaiton	8	_____
2. Wiring splicing	30	_____
3. Soldering	40	_____
4. Taping	7	_____
5. Safety and work habits	15	_____
Total	100	_____

Name _____

Date _____ Grade _____

WIRING A DEAD FRONT PLUG

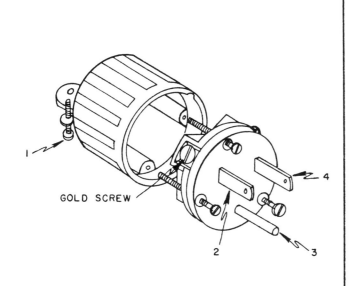

Operational Procedure:

1. Obtain a 3-wire plug cap and a piece of 14 gage 3-wire ground.
2. Disassemble the plug cap.
3. Feed the wire through the plug cap wire clamp and the base of the plug.
4. Rip approximately 1 1/2 inches of the outer insulation off the wire with the cable rippers and cut the insulation off with diagonal pliers.
5. Using the wire strippers, remove 1/2 inch of insulation from the black, white, and green wires.
6. Make a loop in the end of the black, white, and green wires.
7. Properly attach the black, white, and green wires to the proper terminal on the plug cap.
8. Make sure that the connections are neat, tight, and no excess bare wire is exposed.
9. Check with instructor before proceeding to the next step.
10. Attach the dead front cap to the base and tighten dow carefully.

Materials:

1. Industrial quality eye protection
2. Cable ripper
3. Wire strippers
4. Diagonal pliers
5. 1 -- 3 wire dead front plug cap with ground
6. 12", 14-3 SJ wire
7. Long chain nose side cutting pliers

Operation Teaches: Ability to ...

1. Use of cable ripper correctly and safely.
2. Use wire strippers correctly and safely.
3. Use diagonal pliers correctly and safely.
4. Identify and understand what the black, white, and green wires are used for.
5. Identify the parts of a 3-wire plug cap with ground.
6. Wire a 3-wire plug cap.
7. Use safe procedures when wiring.

Evaluation Score Sheet: Points

Item	Possible	Earned
1. Parts properly identified	20	_____
2. Insulation properly removed from wire	15	_____
3. Connection to plug cap made properly	20	_____
4. Neatness	15	_____
5. Plug cap assembled properly	15	_____
6. Safety and work habits	15	_____
Total	100	_____

Part Identification:

1. _____ 3. _____

2. _____ 4. _____

Name _____

Date _____ Grade _____

PORTABLE EXTENSION CORD

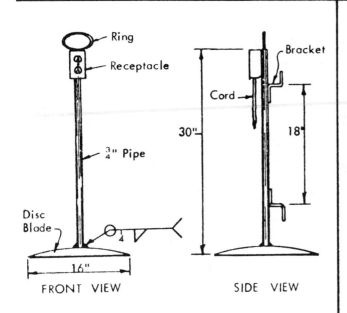

Materials:

1. 1" x 30" black pipe
2. 16" disc blade
3. Grounded duplex receptacle and box
4. Grounded plug (dead front)
5. 2 -- 1/4" x 1" x 10" strap metal
6. 25' 14-3 SJ power cord
7. 1/4" x 15" steel rod
8. 1/4" x 1/2" stove bolt

Operation Teaches: Ability to ...

1. Weld different metals together.
2. Bend rod into a ring.
3. Drill hole in pipe.
4. Tap threads in hole.
5. Heat and bend strap metal.
6. Make electrical connections in circuits.
7. Understand electrical circuits.
8. Understand welding properties of different metals.
9. Tie Underwriters knot.
10. Understand box cable clamps or connectors.
11. Understand proper grounding of electrical circuits.

Operational Procedure:

1. Weld black pipe to disc blade.
2. Bend 1/4 inch rod into a 5 inch diameter circle using a rod bender, or heat it in a forge and draw it into a circle using an anvil and ball peen hammer.
3. Weld ring to top of pipe.
4. Drill 3/16 tap hole 2 inches from top of plate 3.
5. Tap threads into it using 1/4 inch tap.
6. Attach receptacle box to pipe using 1/4 x 1/2 stove bolt.
7. Heat the two 10 inch metal straps, clamp in vise and bend to these dimensions:

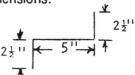

8. Weld these brackets to the pipe using these measurements. Bottom of the upper bracket 5 1/2 inches from top of pipe. Top of lower bracket seven inches from Disc Blade.
9. Push one end of electrical wire through box and attach to receptacle.
10. Clamp cable to box and fasten receptacle to box.
11. Cover receptacle with faceplate.
12. Wrap 25 feet of wire around brackets, push end of wire through plug.
13. Strip back cable covering and wire insulation.
14. Tie underwriters knot and attach wires properly to the plug prongs.

Evaluation Score Sheet:

Item	Possible	Earned
1. Correct measurement	15	
2. Attachment of box brackets, pipe, ring	20	
3. Forming of ring	5	
4. Forming of brackets	5	
5. Tapping threads	5	
6. Connect plug and receptacle correctly	25	
7. Correct ground between plug and receptacle	15	
8. Attitude and work habits	10	
Total	100	

Name _____

Date _____ Grade _____

EXTENSION CORD USING FUSED GROUNDED RECEPTACLE

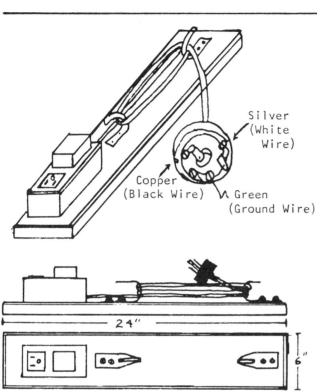

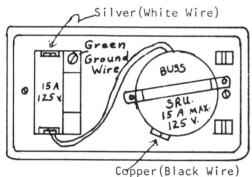

Operational Procedure:

1. Layout the base board.
2. Cut board to proper size and bevel edges.
3. Locate and drill holes for the handy box and conductor mounting brackets.
4. Attach handy box and conductor mounting brackets to base board.
5. Attach the box connector to handy box.
6. Attach conductor to fused receptacle.
7. Attach conductor to grounded cap.
8. Place receptacle in box and attach cover plate.
9. Paint base board and brackets.
10. Insert fusetron for an 8 ampere load.
11. Submit to instructor for evaluation.

Operation Teaches: Ability to ...

1. Understand fusing protable power tools.
2. Understand grounding electric power tools.
3. Follow schematic drawings.
4. Use the drill, drill bits, saws, square, jointer, screwdriver, and wire stripper.
5. Bevel the edges of the base board.
6. Select correct screw size for fastening handy box and brackets to the board.
7. Fasten brackets and handy box to the base board, and attach box conductor to handy box.
8. Strip insulation from the electrical conductor.
9. Wire receptacle and plug correctly.
10. Paint extension cord holder base board and brackets.
11. Select the correct fusetron for the electrical load.

Materials:

1. pine board 1" x 6" x 24"
2. 16' #14-3 rubber covered type SJ conductor
3. Grounded fused receptacle (SRU)
4. Grounded plug (dead front)
5. Fusetron 7 1/2-10 amp range
6. 2 -- mounting brackets, U-type 1/8" x 1/2 x 6
7. 8 -- 3/4" No. 8 round head wood screws
8. 2" x 4" handy box, 1 1/2" deep
9. 3/4" box connector

Name _____

Date _____ Grade _____

Evaluation Score Sheet:

Item	Possible	Earned
1. Correct measurement of baseboard	5	
2. Shaping of conductor mounting brackets	5	
3. Attachment of brackets and box	5	
4. Insulation removed correctly	10	
5. Fused, grounded receptacle, properly wired	25	
6. Cap wired properly	20	
7. Selection of correct fusetron	15	
8. Quality of painting	5	
9. Attitude and work habits	10	
Total	100	

APPENDIX

**OHM'S LAW
and
WATTAGE FORMULAS**

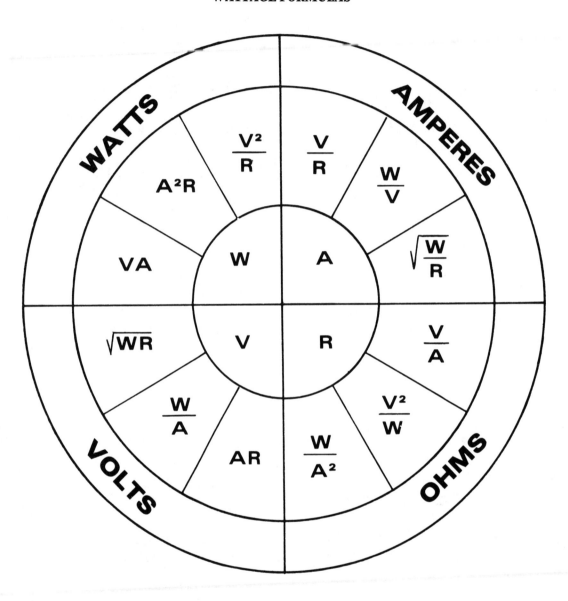

WIRING HANDBOOK TABLES

Table 1

Single-Phase A.C. Motor Currents

Motor Horse-power	115 Volts		230 Volts	
	Full Load (Amps.)	125% Full Load (Amps.)	Full Load (Amps.)	125% Full Load (Amps.)
1/6	4.4	5.5	2.2	2.8
1/4	5.8	7.2	2.9	3.6
1/3	7.2	9.0	3.6	4.5
1/2	9.8	12.2	4.9	6.1
3/4	13.8	17.2	6.9	8.6
1	16	20	8.0	10
1-1/2	20	25	10.0	12.5
2	24	30	12	15
3	34	42	17	21
5	56	70	28	35
7-1/2	—	—	40	50
10	—	—	50	62

Table 2

Minimum Allowable Size of Conductor — Copper up to 200 Amperes, 115-120 Volts, Single Phase, Based on 2% Voltage Drop

Length of Run in Feet. Compare size shown below with size shown to left of double line. Use the larger size.

Load in Amps	In Cable, Conduit, Earth Types R, T, TW	Types RH, RHW, THW	Overhead in Air* Bare & Covered Conductors	30	40	50	60	75	100	125	150	175	200	225	250	275	300	350	400	450	500	550	600	650	700
5	12	12	10	12	12	12	12	12	12	12	10	10	10	10	8	8	8	8	6	6	6	6	4	4	4
7	12	12	10	12	12	12	12	12	12	10	10	8	8	8	8	6	6	6	4	4	4	4	4	3
10	12	12	10	12	12	12	12	10	10	8	8	6	6	6	6	4	4	4	4	3	3	2	2	2
15	12	12	10	12	12	10	10	10	8	6	6	6	4	4	4	4	3	2	2	1	1	0	0	0
20	12	12	10	12	10	10	8	8	6	6	4	4	4	3	3	2	2	1	1	0	0	00	00	00
25	10	10	10	10	10	8	8	6	6	4	4	4	3	3	2	2	1	1	0	00	000	000	000	000
30	10	10	10	10	8	8	6	6	4	4	3	2	2	1	1	1	0	00	00	000	000	000	4/0	4/0
35	8	8	10	10	8	8	6	6	4	4	3	2	2	1	1	0	0	00	000	000	4/0	4/0	4/0	250
40	8	8	10	8	8	6	6	4	4	3	2	2	1	1	0	0	00	00	000	000	4/0	4/0	250	250	300
45	6	8	10	8	8	6	6	4	4	3	2	1	1	0	0	00	00	000	000	4/0	4/0	250	250	300	300
50	6	6	10	8	6	6	4	4	3	2	1	1	0	0	00	00	000	000	4/0	4/0	250	250	300	300	350
60	4	6	8	8	6	4	4	4	2	1	1	0	00	00	000	000	000	4/0	250	250	300	300	350	400	400
70	4	4	8	6	6	4	4	3	2	1	0	00	00	000	000	4/0	4/0	250	300	300	350	400	400	500	500
80	2	4	6	6	4	4	3	2	1	0	00	000	000	4/0	4/0	250	250	300	300	350	400	400	500	500	600
90	2	3	6	6	4	4	3	2	1	0	00	000	000	4/0	4/0	250	250	300	350	400	500	500	500	600	600
100	1	3	6	4	4	3	2	1	0	00	000	000	4/0	4/0	250	250	300	350	400	500	500	500	600	600	700
115	0	2	4	4	4	3	2	1	0	00	000	4/0	4/0	250	300	300	350	400	500	500	600	600	700	700	750
130	00	1	4	4	3	2	1	0	00	000	4/0	4/0	250	300	300	350	400	500	500	600	600	700	750	800	900
150	000	0	2	4	2	1	1	0	000	4/0	4/0	250	300	350	350	400	500	500	600	700	700	800	900	900	1M
175	4/0	00	2	3	2	1	0	00	000	4/0	250	300	350	400	400	500	500	600	700	750	800	900	1M		
200	250	000	1	2	1	0	00	000	4/0	250	300	350	400	500	500	500	600	700	750	900	1M				

Table 3

Minimum Allowable Size of Conductor — Copper up to 200 Amperes, 115-120 Volts, Single Phase, Based on 3% Voltage Drop

Length of Run in Feet. Compare size shown below with size shown to left of double line. Use the larger size.

Load in Amps	In Cable, Conduit, Earth Types R, T, TW	Types RH, RHW, THW	Overhead in Air* Bare & Covered Conductors	30	40	50	60	75	100	125	150	175	200	225	250	275	300	350	400	450	500	550	600	650	700
5	12	12	10	12	12	12	12	12	12	12	12	12	12	10	10	10	10	8	8	8	8	8	6	6	6
7	12	12	10	12	12	12	12	12	12	12	10	10	10	8	8	8	8	6	6	6	6	6	4	4
10	12	12	10	12	12	12	12	12	10	10	8	8	8	8	8	6	6	4	4	4	4	4	4	3
15	12	12	10	12	12	12	12	10	8	8	6	6	6	6	4	4	4	3	3	2	2	2	2
20	12	12	10	12	12	12	10	10	8	6	6	6	4	4	4	3	3	2	2	1	1	0
25	10	10	10	12	12	10	10	8	6	6	6	4	4	4	3	2	2	1	1	0	0	0
30	10	10	10	12	10	10	8	8	6	4	4	4	3	3	2	2	1	1	0	0	00	00
35	8	8	10	12	10	8	8	6	6	4	4	3	2	2	2	1	0	0	00	00	000	000
40	8	8	10	10	10	8	8	6	4	4	3	3	2	2	1	0	0	00	00	000	000	4/0
45	6	8	10	10	8	8	6	4	4	4	3	2	2	1	1	0	00	00	000	000	4/0	4/0
50	6	6	10	10	8	8	6	4	4	3	2	2	1	1	1	0	00	000	000	000	4/0	4/0	250
60	4	6	8	8	8	6	6	4	3	2	2	1	1	0	00	000	000	000	4/0	4/0	250	250	300
70	4	4	8	8	6	6	4	3	2	2	1	0	00	00	000	4/0	4/0	250	250	300	300	300
80	2	4	6	8	6	6	4	4	3	2	1	0	00	00	000	4/0	4/0	250	250	300	300	350
90	2	3	6	8	6	4	4	2	1	1	0	00	00	000	000	4/0	250	250	300	300	350	400	400
100	1	3	6	6	6	4	4	3	2	1	0	0	00	000	000	4/0	250	250	300	350	350	400	400	500
115	0	2	4	6	4	4	3	1	0	00	000	000	4/0	4/0	250	300	300	350	400	500	500	500
130	00	1	4	6	4	4	3	2	0	00	000	4/0	4/0	250	250	300	350	400	400	500	500	600
150	000	0	2	4	4	3	2	1	0	00	000	4/0	4/0	250	250	300	350	400	500	500	500	600	600	700
175	4/0	00	2	4	3	2	1	0	00	000	4/0	250	250	300	300	350	400	500	500	600	600	700	700	750
200	250	000	1	4	3	2	1	0	00	000	4/0	250	250	300	350	350	400	500	500	600	700	700	750	800	900

The above tables were printed courtesy of the Edison Electric Institute and the Food and Energy Council.

*Conductors in overhead in air span must be at least No. 10 for spans up to 50 feet and No. 8 for longer spans.

Table 4

Copper up to 200 Amperes, 115-120 Volts, Single Phase, Based on 4% Voltage Drop

Minimum Allowable Size of Conductor — In Cable, Conduit, Earth / Overhead in Air*

Length of Run in Feet — Compare size shown below with size shown to left of double line. Use the larger size.

Load in Amps	Types R, T, TW	Types RH, RHW, THW	Bare & Covered Conductors	30	40	50	60	75	100	125	150	175	200	225	250	275	300	350	400	450	500	550	600	650	700
5	12	12	10	12	12	12	12	12	12	12	12	12	12	12	12	12	10	10	10	10	8	8	8	8	8
7	12	12	10	12	12	12	12	12	12	12	12	12	12	10	10	10	10	8	8	8	8	6	6	6	6
10	12	12	10	12	12	12	12	12	12	12	10	10	10	10	8	8	8	6	6	6	6	4	4	4	4
15	12	12	10	12	12	12	12	12	10	10	10	8	8	8	6	6	6	6	4	4	4	4	4	3	3
20	12	12	10	12	12	12	12	10	10	8	8	8	6	6	6	6	4	4	4	3	3	2	2	2	2
25	10	10	10	12	12	12	10	10	8	8	6	6	6	6	4	4	4	3	3	2	2	1	1	1	1
30	10	10	10	12	12	10	10	8	6	6	6	4	4	4	4	3	2	2	1	1	1	0	0
35	8	8	10	12	12	10	10	8	8	6	6	4	4	4	3	3	2	1	1	0	0	0	0
40	8	8	10	12	10	10	8	8	6	6	4	4	4	3	3	2	1	1	0	0	0	00	00
45	6	8	10	12	10	8	8	6	6	4	4	4	3	3	2	2	1	1	0	0	00	00	000
50	6	6	10	10	10	8	8	6	6	4	4	4	3	3	2	2	1	1	0	0	00	000	000	000
60	4	6	8	10	8	8	8	6	4	4	4	3	2	2	1	1	1	0	00	00	000	000	000	4/0
70	4	4	8	10	8	8	6	6	4	4	3	2	2	1	1	0	0	00	00	000	000	4/0	4/0	250
80	2	4	6	8	8	6	6	4	3	2	2	1	1	0	0	00	00	00	000	000	4/0	4/0	250	300
90	2	3	6	8	8	6	6	4	3	2	1	1	0	0	00	00	00	000	000	4/0	4/0	250	250	300	300
100	1	3	6	8	6	6	4	4	3	2	1	1	0	0	00	00	000	000	4/0	4/0	250	250	300	300	350
115	0	2	4	8	6	6	4	4	3	2	1	0	0	00	00	000	000	4/0	4/0	250	300	300	350	350	400
130	00	1	4	6	6	4	4	3	2	1	0	0	00	00	000	000	4/0	4/0	250	300	300	350	400	400	500
150	000	0	2	6	4	4	3	3	1	0	00	000	000	4/0	4/0	250	300	300	350	350	400	500	500	500
175	4/0	00	2	6	4	4	3	2	1	0	00	000	4/0	4/0	250	250	300	350	350	400	400	500	500	600	600
200	250	000	1	4	4	3	2	1	0	00	000	000	4/0	250	250	300	350	400	500	500	500	600	600	700

Table 5.

Copper up to 400 Amperes, 230-240 Volts, Single Phase, Based on 2% Voltage Drop

Minimum Allowable Size of Conductor — In Cable, Conduit, Earth / Overhead in Air*

Length of Run in Feet — Compare size shown below with size shown to left of double line. Use the larger size.

Load in Amps	Types R, T, TW	Types RH, RHW, THW	Bare & Covered Conductors	50	60	75	100	125	150	175	200	225	250	275	300	350	400	450	500	550	600	650	700	750	800
5	12	12	10	12	12	12	12	12	12	12	12	12	12	12	10	10	10	10	8	8	8	8	8	6	6
7	12	12	10	12	12	12	12	12	12	12	12	10	10	10	10	8	8	8	8	6	6	6	6	6
10	12	12	10	12	12	12	12	12	10	10	10	8	8	8	8	6	6	6	6	4	4	4	4	4
15	12	12	10	12	12	12	10	10	10	8	8	6	6	6	6	4	4	4	4	4	3	3	3	2
20	12	12	10	12	12	10	10	8	8	8	6	6	6	6	4	4	4	3	3	2	2	2	1	1
25	10	10	10	12	10	10	8	8	6	6	6	4	4	4	3	3	2	2	1	1	1	0	0
30	10	10	10	10	10	8	6	6	6	4	4	4	4	3	2	2	1	1	1	0	0	0	00
35	8	8	10	10	10	8	8	6	6	4	4	4	3	3	2	2	1	1	0	0	00	00	00
40	8	8	10	10	8	8	6	6	4	4	4	3	3	2	2	1	1	0	0	00	00	000	000
45	6	8	10	10	8	6	6	4	4	4	3	3	2	2	1	1	0	0	00	00	000	000	000
50	6	6	10	8	8	6	6	4	4	4	3	3	2	2	1	1	0	0	00	00	000	000	4/0	4/0
60	4	6	8	8	8	6	4	4	3	2	2	1	1	1	0	00	00	000	000	000	4/0	4/0	250
70	4	4	8	8	6	6	4	4	3	2	2	1	0	0	00	00	000	000	4/0	4/0	4/0	250	250	300
80	2	4	6	6	6	4	4	3	2	2	1	1	0	0	00	00	000	000	4/0	4/0	250	250	300	300	300
90	2	3	6	6	6	4	4	3	2	1	1	0	0	00	00	000	000	4/0	4/0	250	250	300	300	300	350
100	1	3	6	6	4	4	3	2	1	1	0	0	00	00	000	000	4/0	4/0	250	250	300	300	350	350	400
115	0	2	4	6	4	4	3	2	1	0	0	00	00	000	000	4/0	4/0	250	300	300	350	350	400	400	500
130	00	1	4	4	4	3	2	1	0	0	00	00	000	000	4/0	4/0	250	300	300	350	400	400	500	500
150	000	0	2	4	4	3	1	0	0	00	000	000	4/0	4/0	250	250	300	350	350	400	500	500	500	600
175	4/0	00	2	4	3	2	1	0	00	000	000	4/0	4/0	250	250	300	350	400	400	500	500	600	600	600	700
200	250	000	1	3	2	1	0	00	000	000	4/0	4/0	250	250	300	350	400	500	500	500	600	600	700	700	750
225	300	4/0	0	3	2	1	0	00	000	4/0	4/0	250	300	300	350	400	500	500	600	600	700	700	750	800	900
250	350	250	00	2	1	0	00	4/0	4/0	250	250	300	350	350	400	500	500	600	700	700	750	800	900	1M
275	400	300	00	2	1	0	00	000	4/0	250	250	300	350	400	500	500	600	700	700	800	900	900	1M
300	500	350	000	1	1	0	000	4/0	4/0	250	300	350	350	400	500	500	600	700	700	800	900	900	1M
325	600	400	4/0	1	0	00	000	4/0	250	300	300	350	400	500	500	600	700	750	900	900	1M
350	600	500	4/0	1	0	00	4/0	250	300	300	400	400	500	500	600	700	750	800	900	1M
375	700	500	250	0	0	00	4/0	250	300	350	400	500	500	600	600	700	800	900	1M
400	750	600	250	0	00	000	4/0	250	300	350	400	500	500	600	700	750	900	1M

*Conductors in overhead in air spans must be at least No. 10 for spans up to 50 feet and No. 8 for longer spans.

The above tables were printed courtesy of the Edison Electric Institute and the Food and Energy Council.

Table 6

GRAPHICAL ELECTRICAL SYMBOLS FOR GENERAL FARM WIRING PLANS

These symbols have been extracted from American National Standards Institute (ANSI) Standard Y32.9 Graphic Electrical Wiring Symbols for Architectural and Electrical Layout Drawings; and from ANSI Standard Y32.2-Graphic Symbols for Electrical Diagrams. The numbers in parenthesis ( ) are ANSI-Y32.2 Symbol numbers. Other numbers are for ANSI Y32.9 Symbols.

1.0 Lighting Outlets
Ceiling
- 1.1 Surface or pendant incandescent or similar lamp fixture
- 1.3 Surface or pendant individual fluorescent fixture
- 1.5 Surface or pendant continuous-row fluorescent fixture
- 1.10 B — Blanked outlet
- 1.11 J — Junction box

Wall
To indicate wall installation of above outlets, place circle near wall and connect with line as shown

2.0 Receptacle Outlets
- 2.1 Single receptacle outlet
- 2.2 Duplex receptacle outlet
- 2.3 Triplex receptacle outlet
- 2.5 Duplex receptacle outlet -- Split wired
- 2.7 Single special-purpose receptacle outlet *
- 2.8 Duplex special-purpose receptacle outlet *
- 2.9 R — Range outlet
- 2.12 C — Clock hanger receptacle

3.0 Switch Outlets
- 3.1 S — Single-pole switch (SPST)
- 3.2 S_2 — Double-pole switch (DPST)
- 3.3 S_3 — Three-way switch (SPDT)
- 3.4 S_4 — Four-way switch (DPDT)
- 3.6 S_P — Switch and pilot lamp
- 3.9 Switch and single receptacle
- 3.10 Switch and double receptacle
- 3.11 S_d — Door switch
- 3.12 ST — Time switch
- 3.15 S — Ceiling pull switch

Power, Fusing, Ground
- (46.3) MOT — Electric motor
- (46.2) GEN — Electric generator
- (86.1) Power transformer
- (48) WH — Electric watthour meter
- (11.1) Circuit breaker
- (36) Fusible element
- (13.1) Ground

6.0 Panelboards, Switchboards and Related Equipment
- 6.2 Surface-mounted panelboard and cabinet
- 6.7 Motor or other power controller *
- 6.8 Externally operated disconnection switch *
- 6.9 Combination controller and disconnection means *

8.0 Remote Control Stations for Motors or Other Equipment
- 8.1 Pushbutton station
- 8.2 F — Float switch -- Mechanical
- 8.3 L — Limit switch -- Mechanical
- 8.4 P — Pneumatic switch -- Mechanical
- 8.5 T — Thermostat

Low-Voltage and Remote-Control Switching Systems
- 1.12 L — Outlet controlled by low-voltage switching when relay is installed in outlet box
- 3.7 S_l — Switch for low-voltage switching system
- 3.8 S_{lm} — Master switch for low-voltage switching system

* Use number or letter either within the symbol or as a subscript alongside the symbol keyed to explanation in the drawing list of symbols to indicate type of receptacle or usage. Use supplemental symbol schedule, lower left:

Special Identification of Outlets: Supplemental Symbol Schedule

WP	Weather proof	DT	Dust tight
VT	Vapor tight	EP	Explosion proof
WT	Water tight	G	Grounded
RT	Rain tight	R	Recessed

- The electrical symbols appearing on this page provide a key to the wiring symbols used in the drawings appearing in this Handbook.
- Where all or a majority of receptacles in an installation are to be of the grounding type, the upper case letter abbreviated notation may be omitted and the types of receptacles required noted in the drawing list of symbols and/or in the specifications. When this is done, any non-grounding receptacles may be so identified by notation at the outlet location.

The above tables were printed courtesy of the Edison Electric Institute and the Food and Energy Council

Table 7

GRAPHICAL ELECTRICAL SYMBOLS FOR SPECIFIC FARMSTEAD EQUIPMENT WIRING PLANS

(All symbols conform to American National Standards Institute, Standard, ANSI Y 32.2, 1970. Numbers preceding the symbols **do not** coincide with ANSI numbers.)

#	Symbol	#	Symbol
1	Resistor	21	Time delay closing, N.O.
2	Variable resistor	22	Time delay opening, N.O.
3	Potentiometer	23	Time delay closing, N.C.
4	Capacitor	24	N.O. relay contacts
5	Coil or inductor	25	N.C. relay contacts
6	Grounded terminal	26	SPDT relay contacts
7	Battery	27	Make-before-break SPDT relay contacts
8	Connection	28	DPDT relay contacts
9	No connection	29	Thermostat
10	Thermo-mechanical transducer	30	Thermal time delay relay, N.O. contacts
11	Normally open (N.O.), pushbutton switch	31	Overload protector with N.C. contacts
12	Normally closed (N.C.), pushbutton switch	32	Auto-transformer, adjustable
13	Single-pole, double-throw pushbutton switch	33	Isolation transformer
14	Single-pole, single-throw switch	34	Motor
15	Single-pole, double-throw switch (SPDT)	35	Diode
16 a	Circuit breaker	36	Zener diode
b	Fuse or fusible element	37	SCR
17	Selector switches	38	Triac
18	N.O. limit switch	39	PNP transistor
19	N.C. limit switch	40	NPN transistor
20	Time delay opening, N.C.		

The above tables were printed courtesy of the Edison Electric Institute and the Food and Energy Council.

TABLE 8. Approximate Power Requirements of Common Electrical Equipment and Small Appliances Found Around the Home.

EQUIPMENT	WATTS
Air conditioner central, electric	3000-9000
Air conditioner, room	800-3000
Blender	200-300
Blanket, electric	150-200
Clock	2-3
Coffee maker	500-1000
Corn popper	450-600
Crock-pot, slow cooker	150-300
Dishwasher	600-1200
Dryer, clothes	4000-5000
Entertainment center no TV	30-100
Fan, 8-12 inch portable	40-80
Fan, kitchen vent	100
Freezer, household	300-500
Fryer, deep fat	1200-1650
Frying pan	1000-1200
Furnace, oil fired (fan and burner)	600-800
Garbage disposal	300-600
Heater, permanent wall type	1000-2300
Heater, portable household	1000-1500
Heater, water	2000-5000
Hot plate, per burner	600-1000
Ironer	1200-1500
Iron, hand	660-1200
Knife, electric	100
Lamps, fluorescent	15-60
Lamps, incandescent	15-1000
Micro-wave oven	700-1500
Mixer, food	120-250
Motors, less than 1 hp	1200/hp
1 hp and above	1000/hp
Polisher, floor	250
Projector, movie or slide	300-1000
Radio	10-80
Range, oven only	4000-5000
Range, top only	4000-6000
Razor	8-12
Refrigerator, household	200-400
Roaster	1200-1650
Sewing machine	60-90
Soldering iron	100-300
Television	200-400
Toaster	600-1200
Vacuum cleaner	250-800
Waffle iron	600-1000
Washer, automatic	600-800
Washing machine	350-550

DEFINITION OF ELECTRICAL TERMS

Alternating Current (A-C): A current which reverses its direction of flow.

Alternator: A device for changing mechanical energy, usually rotating, into alternating current electricity.

Ammeter: An instrument used to measure the rate of flow of electrical current in a circuit.

Ampacity: The safe carrying capacity of an electrical wire, in amperes.

Ampere: The rate of flow of electricity. Current flow means the same in electricity.

Apparent Wattage: The wattage as determined by the use of a voltmeter and ammeter to arrive at watts being used by an induction load such as an electric motor. It is a false reading and higher than what it should be. See True Wattage.

AWG: American Wire Gage, a set of standards used to size conductor wires.

Bonding: A term meaning the same as grounding, see Grounding.

Blown Fuse: A fuse which has been overloaded and its element severed to stop the flow of current in the circuit. The entire fuse or at least the element would need to be replaced to make it operative.

Bronze Color: A color synonymous with brass. Both colors denote **hot** current-carrying conductor points of attachment.

Cable: Individual insulated solid conductors, grouped with insulation around them; or, a number of stranded conductors grouped and insulated, and then these are grouped and insulated into a cable. See Power Cord.

Cap: Sometimes called **plug-in** or **male plug**, but is the device that is wired to a power cord and inserted into a receptacle. Caps come in many different configurations, ampere and voltage ratings.

Cartridge Fuse: A single-element fuse that is cylindrically shaped. Those rated more than 60 amperes have a knife-blade terminal attached to each end.

Circuit: A path over which electrical current will flow.

Circuit Breaker: A switch-type overload current protector that opens automatically when excessive current flows and can be manually reset. Special forms of circuit breakers will reset automatically.

Color Coding: A system established by NEC for safety purposes to have a readily acceptable and a universal understanding of coding by colors of the various parts of the electrical wiring system and fixtures for attaching to the system.

Conductor: Materials allowing free electrons to move through it easily. Copper and aluminum are the common metals used for electrical conductors. Conductors have low resistances to electron flow.

Conduit: A tube or pipe through which electrical wires are run to give the wires protection from mechanical damage and weathering. Conduit serves as the grounding system, and in lieu of the green or bare conductors on many electrical systems.

Continuity: Going on or extending without a break or interruption. A continuous state or unbroken.

Continuity Tester: An instrument having its own power source (usually battery) and a resistor (lamp, bell, buzzer) that when connected in series to conductors current will flow and the resistor will indicate the flowing of current and **continuity**. Continuity tester should be placed only on disconnected parts of a wiring system. An ohmmeter may be used as a continuity tester.

Current: The amount of amperes flowing through a conductor.

Current-carrying Conductor: A conductor which under normal use will carry current. The voltage potential of a conductor may be dangerous (as with **hot** conductors) or safe (as with **neutral** conductors).

Current Overload Protection: Devices, usually called fuses, designed to open the circuit when a predetermined current has been exceeded.

Cycle: A complete flow of current in each direction as in an AC system.

Delta Connection: A three-phase transformer connection so that the phases form a triangle. 120 volts can be taken from each corner of the triangle and a center top, and 240 volts taken from any two corners of the triangle.

Direct Current (D-C): A current that does not change its direction of flow or alternate.

DPDT: Double-pole, double-throw device, when specially wired is used as a four-way switch, or as a switch to reverse certain motors.

DPST: Double-pole, single-throw switch device used in entrance panels as the service disconnect switch to control each **hot** side of the 120/240 volt single-phase service at the same time. Also commonly used as a switch on 240 volt equipment to control this flow of current on both **hot** conductors.

Dual Element Fuse: A time-delay, Edison base, plug-type fuse. See Fusetron.

Earthen Ground: The act of placing a conductor or piece of equipment in direct contact with a metallic ground stake, water pipe or some other device which makes good contact with the ground.

Edison Base: The screw base configuration of common 120 volt light bulbs, lamp receptacles and plug-type fuses used ordinarily in homes.

Electric Motors: A device for changing electric energy into mechanical energy.

Electromotive Force [Emf]: The pressure or force in electricity. The same as voltage.

Energized: The act of connecting to electrical power or causing voltage to be available in a circuit. Example, a switch turned **on** will **energize** the lamp circuit the switch controls.

Entrance Panel: The box or panel from which all circuits in a building originate and branch out.

Equipment Grounding: The use of a separate conductor or metallic conduit from frame or case of electrically powered equipment to an earthern ground for safety purposes. See Grounding and Ground Rod.

Fault: A break, leak or opening in a circuit.

Four-way Switch: A switch used in conjunction with three-way switches to control circuit resistor(s) from three or more locations. See DPDT.

Fractional Horsepower (hp) Motors: Those motors whose related horsepower is less than one, for example, 3/4, 1/2, 1/3 and 1/4 and other small motors. See Horsepower.

Frequency: The number of cycles per second. In North America electrical power is normally 60 cycles per second.

Fuse: A device designed to open the circuit when a specific current has been exceeded. Non-time delay fuses open more quickly than do time-delay types, which can also withstand normal overload currents above their ratings for several seconds before opening the circuit.

Fusetron: A brand-name commonly used for Edison base, dual element (time-delay) fuses. See Dual Element.

Fustat: A trade-name commonly used for non-tamperable fuse.

Generator: A device for changing mechanical energy, usually rotating, into alternating current or direct current electricity.

GFCI: Ground-fault circuit interruptor, a safety device that can sense small amounts of current flowing to ground and quickly break or disconnect the **hot** conductor source. Most styles are overload current protection devices having a specific ampacity rating and can protect equipment, and also protect people against electrical shock hazards.

Green Color: Color of attachment points on fixtures, such as receptacles, to which green or bare color noncurrent-carrying conductors are connected for grounding and safety purposes.

Ground: Any part of the wiring system that has continuity or bonding to earth. The same as grounded. See Neutral Conductor and Grounding.

Ground Rod: A metallic stake or rod driven into earth and connected with a conductor wire to the entrance panel to: (1) serve as a ground for the neutral current-carrying conductor part of the wiring circuit, and (2) provide grounding of all metallic boxes and conduit in the wiring system.

Grounded: A conductor, box or any other device that is in continuity with earth. See Neutral Conductor and Ground Rod.

Grounding: The third-wire, a noncurrent-carrying conductor, of a 120 volt wiring system that provides a separate path for accidental voltage potential to be sent safely to the ground in case of wiring system malfunction, rather than through humans. Another similar term is bonding. See Ground Rod and Neutral Conductor.

Hot Conductor: A conductor that is carrying voltages of sufficient potential to cause serious electrical shock, injury or death if a person makes contact. The **hot** conductor is identified by a color other than white, green or bare; and, is commonly black or red.

Horsepower (hp): A unit of power equal to 746 theoretical watts in electricity, and 33,000 pound-feet per minute in mechanical energy. Rules of thumb for estimating watts per horsepower is 1,000 watts/hp for one hp motors and over, and 1,200 watts/hp for fractional horsepower motors.

Identified Colors: There are only three identified colors, namely **white** as a neutral grounded current-carrying conductor; and **bare** and **green** as grounding noncurrent-carrying conductors for safety purposes. All other colors, such as black, red, yellow, orange and brown are non-identified colors and considered **hot** current-carrying conductors.

Kilowatt: A unit of power equal to 1000 watts.

Kilowatt-hour Meter: An instrument designed to measure and record the use of electrical power or energy. The device used to charge customers for the amount of electrical power consumed.

Lamp: A resistor device designed to produce light.

Light Bulb: Same as lamp. See Lamp.

Line: Any electrical circuit for power or control.

Load: Anything that is connected to a circuit and consumes power, such as a lamp, a toaster or a motor.

National Electrical Code: A set of minimum standards and recommendations for **safe** wiring written by U.S. fire insurance companies. The **Code** is often, but not always mandated by law, enforced and policed through regulatory officers hired by governing bodies such as cities, towns and counties.

Neutral Conductor: A current-carrying conductor, identified as **white** in color, that is grounded to earth and has little or no electrical voltage potential. Neutral, ground and grounded are similar terms.

Non-tamperable Fuse: A dual-element fuse which has a special base whereby a larger ampacity fuse can not be placed into a special Edison base adapter shell which when placed into an entrance panel is not replaceable. This system is used for safety reasons. Type S is the common non-tamperable fuse.

NM: A type of nonmetallic sheathed cable designed for only indoor and permanently dry locations. The cable contains two or more T or TW wires, these are wrapped with tough paper that has been treated with fire retardant compounds. A plastic like jacket surrounds all conductors into one cable. NM **can not** be used for agricultural or industrial wiring and is used most exclusively for residence wiring.

NMC: A high quality nonmetallic sheath cable similar to type NM, except better quality plastics to retard moisture is used without paper being used. The insulated individual conductors are embedded in a solid sheath of plastic material and sometimes a glass over-wrap is used around the insulated conductors to allow for give and take of the cable and ease of jacket removal during wiring preparation.

Noncurrent-carrying Conductors: A conductor which under normal circumstances should not be carrying current, but will under unsafe or fault conditions. The grounding conductor, or third wire, on a 120 volt system is a good example.

No-load: A situation when an electric motor is running but is doing no useful work.

Ohm: A measure or unit of resistance. One ampere will flow through one ohm at one volt of electromotive force.

Ohmmeter: An instrument that has its own source of power and used to measure resistance in ohms on disconnected electrical resistors.

Over Current: A current greater than the design of a device or circuit.

Overload: A power dissipation greater than that for which the circuit or parts of a circuit are designed.

Parallel Circuit: A hookup of two noninterrupting conductors to which two or more resistors are individually placed across.

Plug Fuse: A fuse with an Edison base, similar to the screw portion of an ordinary 120 volt light bulb.

Plug-in: Sometimes called male plug or cap. See definition of Cap.

Polarity: A method of wire identification and fixture design for safety. In practical wiring design, it is to separately maintain the continuity of the **neutral** and also the **hot**. At least two wires are always present for an electrical device to be polarized.

Power Cord: A synonymous name is portable or extension cord. A flexible type of electrical cable, made from many individual strands (usually copper) and these stranded wires are covered with rubber-like or plastic coatings. Two or more of these are then grouped together and covered with another rubber-like outside coating. Power cords are classified by a lettering system indicating intended service applications. For example, type SJ is a medium duty portable cord designed for household appliances, office machines and small motors used inside. Type SJTO, another medium duty cord, is of better quality, has superior oil resistance and recommended for many non-industrial shop applications. Type SO is a heavy duty cord and suitable for nearly all industrial uses. Lampcord, types SPT (PVC insulation) and SP (rubber insulation), are interior applications intended for lamps, clocks, radio and other light duty uses.

Power Factor: A relationship between true wattage and apparent wattage. A factor needed to correct the basic wattage formula when the wattage of an induction load, such as an electric motor is being determined by an ammeter and voltmeter. Power factor on an electric motor is equal to true wattage divided by apparent wattage.

Receptacle: The part of the electrical system mounted in an electrical box and connected to conductors. Receptacles are used by plugging power cord caps into them. Receptacles come in many different configurations, ampere and voltage ratings.

Resistance: That which opposes the flow of electricity. The amount of resistance is measured in ohms.

Resistor: A device that retards the flow of electrical current and converts electrical energy into light, heat or magnetic forces as with lamps, heaters or motors.

Schematic: A sketch, plan or graph that helps to explain or illustrate something by outlining its parts in the form of a drawing or visual configuration.

SE: Service Entrance (SE) cable designed to be used from building point of attachment to entrance panel. This is normally down the side of a building and through the building wall. The SE cable is usually two **hot** insulated stranded conductors with a braided sheath insulated conductor used for ground and surrounds the two **hot** conductors.

Series Circuit: An electrical circuit in which two or more resistors are placed end to end or in line across two conductors.

Service Entrance: The part of the electrical system that transfers electricity from the power supplier's transformer to the service entrance panel.

Short-circuit: A connection between two lines or energized parts of such low resistance that excessive current flows, as in the case when a **hot** and **neutral** makes contact.

Silver Color: A color identified as neutral, ground, or grounded, but not grounding or bonding. Silver color terminals and screws are attached to white neutral conductors.

Single-pole Switch: Called by other names as SPST, two-way switch, toggle switch, and simple switch, and on-off switch. See SPST.

Single-phase: The alternation of an electrical current from a potential of zero, to a position change, back to zero, to a negative charge, and back to zero again during a cycle.

Single Element Fuse: A fusing device that has a single fusible link. Common configurations are Edison base and cartridge styles. Often called by other names, such as **standard** or **non-time delay.**

Solderless Connector: A mechanical splicing device, such as a wire nut or split-bolt connector, used to join two or more electrical wires without soldering.

SPDT: Single-pole, double-throw switch device, when specially wired is used as a three-way switch.

SPST: Single-pole, single-throw device commonly used as a simple toggle switch. It has capabilities of stopping the current flow (OFF) or allowing the current to flow (ON).

Standard Fuse: A single element, non-time delay fuse. See Single Element Fuse.

Stranded: When a number of individual wires make up a total conductor.

Sub-station: A group of transformers placed at a common site.

Switch: A device for opening, closing, or making some alternate connection in a circuit.

Three-phase: The alternations of three, single-phase electrical currents overlapping each other by one-third during a cycle.

Three-way Switch: A type of switch commonly used in pairs to control circuit resistor(s) from two different locations. See SPDT.

Time-delay Fuse: Any fuse having capability of carrying about twice its rated ampacity for several seconds, but will **blow** or **trip** at its rated ampacity under continuous overload current conditions. Sometimes called **time-lag.**

Third-wire: The noncurrent-carrying conductor of the typical 120 volt wiring circuit. It is the same as grounding. See Grounding.

Tinning: A process of soldering whereby insulation is stripped from stranded wire conductors, the wires dipped in soldering paste, heated with soldering iron or gun, and solder applied so that all individual strands are joined as one. Tinning greatly improves the strength of fastening power cord conductors to all terminals and caps.

TPDT: Triple-pole, double-throw switch device uses to reverse the directional rotation of certain types of electrical motors. Not commonly used in ordinary home and farm wiring applications.

TPST: Triple-pole, single-throw switch device used in three-phase entrance panel as the service disconnect switch to control all three **hot** conductors. Also commonly used as a switch on three-phase power equipment to control the flow of current on the three **hot** conductors.

Transformer: A device for increasing or decreasing the voltage and the current as electricity passes through it.

Triplex: A type of overhead in air service entrance cable which has two insulated current-carrying conductors spiralling around a strong central ground cable made from steel and aluminum to help support the entire cable.

Tripped Fuse: Usually makes reference to a circuit-breaker type current overload protection device that has been opened because of excessive load. It can be reset from the **tripped** position and readied for operation again without part replacement.

True Wattage: The wattage as determined by the use of a kilowatt-hour meter as the actual watts being used in an induction load, such as an electric motor. It is called **true** because it is correct, as contrasted to apparent wattage on electric motor measurement. See Apparent Wattage.

TW: Single insulated conductors (no cable) made from either stranded or solid copper. Used for general purpose wiring in conduit and raceways.

UF: A PVC covered cable used as a underground feeder (UF). It may be used as a feeder or branch circuit, conductor in direct burial situations, but can not be used as a main service entrance cable drop or lateral. It may be used as interior wiring especially in high moisture areas.

USE: Underground Service Entrance (USE) cable is a high quality cable designed to be buried in earth and make the run from transformer to entrance panels in various buildings. USE cables are solid insulated conductors, either copper or alumnium.

Voltage: The potential force or pressure of electricity.

Voltage Drop: Loss of potential voltage that occurs when current flows through a resistor.

Voltmeter: An instrument designed to measure the electrical potential or voltage between two points in a circuit.

Watt: A unit of electrical power or the rate of using electrical energy. A watt is equal to one ampere at one volt.

Wattage: The rating of a resistor that will consume a certain amount of electrical energy when used at a specific voltage.

Wattmeter: An instrument that measures power in watts.

WP: A weather proof (WP) single insulated solid copper conductors (not cable) to make a run overhead in air from transformer to entrance panels in various buildings.

Wye Connection: A transformer connection with the three windings going outward from a central neutral ground point. 120 volts can be taken from each of the three external points, and 208 volts from any two of the three terminals.